Combating Corrosion: Strategies to Protect Our Infrastructure and Environment

Cronin

Contents

Chapter 1

Introduction

1.1 Motivation

1.1.1 Corrosion

Corrosion is the deterioration of a material as a result of a electrochemical reaction with its environment. The nature of the chemical reaction involves the transfer of electrons at the metal surface, i.e. a change in the oxidation state of the metal. The oxidation reaction is accompanied by its respective reduction or cathodic reaction via the transfer of ions through the electrolyte. In other words, both half cell reactions occur simultaneously in order for the corrosive process to take place. There are eight common forms of corrosion, namely: uniform, galvanic, pitting, crevice, intergranular, dealloying, erosion and environmentally assisted cracking.[1,2] Each of the eight forms of corrosion can be more broadly categorized into either an uniform or a localized corrosion. In the uniform corrosion, no preference of the metal surfaces are attacked more than other area. This results in the thinning of the overall substrate uniformly until the eventual failure of the material. Although uniform corrosion gives the appearance that the corrosion occurs at all surface at any given time. However, localized anodic and cathodic spots exist but move randomly on the surface of the metal. On the other hand, localized corrosion is defined as having fixed positions of the local anodic and cathodic sites. Therefore, corrosion only occurs at those locations. For example, pitting of the metal is the attack of a spot at which the protective film breaks down locally. Thus, only the spot where the bare metal is in contact with its environment experience the corrosion attack.[2,3]

The negative consequence of corrosion has impacted our society in the form of finance, safety, environment, as well as the conservation of resources. The average annual direct cost of corrosion in Canada is estimated to be approximately $41 billion, which does not include the indirect contribution which often refers to as the social costs.[4] For comparison, in most industrialized countries, the total cost of corrosion ranges somewhere between 1 to 5% of the gross domestic product (GDP) averaging around 3.5%.[4] The United States alone, has the direct corrosion cost at approximately 276 billion dollars (or 3.1% GDP) per annum.[5] To put in perspective, the current estimate per capita direct cost of corrosion per citizen in a developed country is approximately $1200 a year. However, 25 to 30% of the costs of corrosion could be avoided if optimum corrosion management practices are employed.[4]

Aside from the financial aspect of corrosion, the safety concern is certainly a much pressing problem with the infamous incident of the Aloha airlines as an example, where multiple people died due to airplane structural failure caused by corrosion.[2,6] Finally, the environmental impact of corrosion involves both an increase in consumed energy in a form of an extractive metallurgy in the reverse direction as well as potential hazard leakage and exposure due to the deterioration of storage and containment of waste materials that are harmful to human. As an example, an incident that took place in Bhopal, India in 1984 where a pesticide plant released 50 tonnes of methyl isocyante gas due to corroded pipe ultimately took the life of more than 20,000 people; an additional 50,000 people suffered from long term illness.[7]

1.1.2   Corrosion Inhibition

In the field of corrosion inhibition, there are various way of minimizing corrosion taking place by either breaking the electronic/ionic contact or slowing both the anodic and cathodic partial reactions. The former can be achieved with electrical isolation approaches or coatings while the latter can be achieved with the addition of inhibitors. In practice, both approaches are often used in conjunction. For example, in natural waters, the corrosion cell process is chiefly dependent upon the oxygen reduction reaction. Therefore, the common approach is often to increase the oxidizing power of solution to promote passivisation on the metal surface while employ oxygen scavenger molecules such as sodium sulfite or hydrazine for the extraction of oxygen molecules. In most cases, additional chemical inhibitors are added to reinforce the passivity.[8]

Surfactants are generally a class of organic compounds that consist of a hydrophilic head and a hydrophobic tail. They can be broadly categorized into four different types based on their head group charges: anionic, cationic, nonionic and zwitterionic.[9] In the context of liquid–solid interfaces, surfactant molecules aggregate such that the chemical interaction with the liquid phase dictates the formation of micelles (in aqueous phase) or reverse micelles (in organic phase). In general, when the critical micelle concentration (CMC) is exceeded and/or another medium is placed in proximity, surfactants preferentially adsorb onto the nearby surface via physical or chemical adsorption. Such surface interaction is highly dependent on the CMC, solvent polarity, temperature, pressure, and pH.[10]

The utilization of chemical corrosion inhibitors are ubiquitous in various industries to combat the effects of corrosion on the structural integrity of metallic substrates.[9, 11–14] Surfactants as corrosion inhibitors are environmentally friendly, have high inhibition efficient, low cost and are easy to produce relative to other corrosion inhibition methods.[9] Most surfactants as corrosion inhibitors work by forming a protective layer on top of the metallic surface to prevent the electrochemical process of corrosion. Its mechanism depends on the formation of a mono- or multilayer, where the protective nature of the surface layer depends on factors such as the interaction between surfactant and metallic substrate, electrode potentials, concentration of the surfactant, and temperature.[10] In an acidic environment, iron in steel reacts with oxygen in water (or air at high temperatures) to form iron oxide. The iron oxide then reacts with excess of hydronium (at low pH) to form dissolved iron ions and water.[10]

In recent years, alkyl phosphate esters have been adopted to combat the effect of iron/steel corrosion in acidic conditions in the petroleum industry. Their effectiveness have been demonstrated by the commercial availability of numerous blend of inhibitors containing alkyl phosphate esters. However, detail studies of how these species structure at the surface are lacking. A better understanding of the molecular details of phosphate ester–iron interactions is required in order to more effectively combat energy, environmental, and financial loss. Furthermore, understanding how these surfactants are structured on metal surfaces can potentially accelerate the design of improved inhibitors to further minimize corrosion.

1.2   Second-order Nonlinear Optics

1.2.1   Basic concepts

Nonlinear optics is the study of the interaction between light and matter and its occurrence as a result of the modification of the optical properties of the system in the presence of an external electric field. This interaction can be described as the sum of the molecular electric dipole moment per unit volume or the macroscopic polarization (P), which can be described as

$$P = \varepsilon_o(\chi^{(1)}E + \chi^{(2)}E^2 + \chi^{(3)}E^3 + \cdots) \tag{1.1}$$

,where ε_o is the vacuum permittivity, and $\chi^{(n)}$ is the electric susceptibility.

The nonlinearity in nonlinear optics comes from the induced polarization deviating from the linear dependence of the electric field strength. In the three-wave mixing scheme, where two input fields are used to generate the third field, the second order nonlinear polarization is proportional to the second term in Eqs. 1.1. Therefore, to focus on the second order nonlinear optics, the expression can be shorten to

$$P^{(2)} = \varepsilon_o \chi^{(2)} E_1 E_2 \tag{1.2}$$

where $\chi^{(2)}$, the second order nonlinear susceptibility, is a third rank tensor with 27 elements in the Cartesian coordinates describing the relationship between the induced second order polarization and the incident field. The second order process is inherently forbidden in a centrosymmertic environment with an inversion center. This can be seen if we perform an inversion operation on the coordinate system in Eqs. 1.2 where vector quantities such as $P^{(2)}$, E_1, and E_2 change sign which leads to

$$-P^{(2)} = \varepsilon_o \chi^{(2)} (-E_1)(-E_2). \tag{1.3}$$

Since both equations above are to be true, therefore $\chi^{(2)}$ must be zero. This rule however, only takes into account the electric dipole contribution ignoring the higher order contributions. Such approximation is known as the electric dipole approximation which forms the basis of its surface specificity. The higher order multipolar contributions are often insignificant compare to the electric dipole response. Therefore, such approximation is generally valid in many cases.

Second-order nonlinear, therefore can be an useful tool for vibrational studies at interfaces when the same species are present in one or both of the adjacent bulk phases. This is the result of the symmetry requirement for non-zero achiral second order susceptibility tensor elements $\chi^{(2)}_{ijk}$ where i, j and k are Cartesian coordinates. Under the electric dipole approximation, and in the absence of appreciable surface charges, signal originating from $\chi^{(2)}$ processes such as sum-frequency generation (SFG) are observed only in the absence of inversion symmetry.[15, 16] That is, there can be no macroscopic inversion center that creates a point $(x, y, z) \rightarrow (-x, -y, -z)$. This occurs to some extent for purely mathematical reasons as described above. For example, at the solid-liquid interface, if we imagine a Fresnel-type interface where a semi-infinite solid phase meets a semi-infinite liquid phase, no inversion centers exist at the plane $z = 0$. If the solid phase is amorphous, there is an effective inversion center in the isotropic bulk, as there is in the bulk liquid phase. If a methyl symmetric stretch is observed at the interface, we therefore know that no methyl groups in the bulk solid phase have contributed to the measured response. In addition to this geometric symmetry-breaking, the anisotropic forces acting on a methyl group with a bulk solid phase located at $z < 0$ and a bulk liquid phase at $z > 0$ likely results in a preferred organization of the chemical functional groups. This creates an additional symmetry breaking in a chemical sense. As a final note, if there are species in solution that contain methyl groups, they will also not contribute to the vibrational $\chi^{(2)}$ spectrum since their orientational average is isotropic unless they interact with (are adsorbed on) the surface.

In order to achieve the second order nonlinear process, the spatial and temporal overlaps of the two input fields are essential. In an example of the sum frequency generation (SFG), the generation of the third field, namely the sum frequency field ($\omega_{SFG} = \omega_1 + \omega_2$) at an angle θ_{SFG} which can be calculated from the conservation of momentum as

$$n_{SFG}\,\omega_{SFG}\,\theta_{SFG} = n_1\,\omega_1\,\theta_1 + n_2\,\omega_2\,\theta_2 \qquad (1.4)$$

where n is the refractive index of the medium, ω is the frequency and θ is the angle of incidence.

1.2.2 Nonlinear vibrational Spectroscopy

The origins of the utilization of second-order nonlinear optics as a spectroscopic technique can be traced back to the first observation and demonstration of the nonlinear-optical phenomena by Franken et al. in 1961 where the University of Michigan based group demonstrated the frequency doubling of a 694 nm ruby laser light source from a piece of quartz.[17] Such realization of the phenomena has accelerated the interest from the field of laser spectroscopy in the rapidly growing research of interfacial studies owing to the distinct advantage because of its interfacial specificity in comparison to linear optical based technique.[18–23] The interfacial specificity and sensitivity has enabled numerous research areas including physical chemistry and biochemistry that previously were not able to realized due to the inaccessibility of the signal generated from the source of interest.

For elaboration, given an example of an organic liquid-substrate interface, Linear optical based vibrational technique such as IR absorbance measurement is not suitable to examine the interfacial molecules due to the signal can arise from both the bulk phase as well as the interface. Given the relative numbers of molecules present in bulk liquid in comparison to interfacial molecules, the signal inevitably will be swamped by the bulk signal contribution. On the other hand, nonlinear vibrational spectroscopy is capable of probing only the interfacial molecules given that the bulk liquid phase is centrosymmetric in nature (For obvious reason, this will exclude solution such as chiral molecules in liquid phase). Another advantage of nonlinear optical based vibrational spectroscopic technique over its linear counterpart is that of the detection limit of the organic monolayer. Given an example of a monolayer structured on metal surface without the present of the liquid phase. Traditional IR reflection absorption measurement would seen to be an accessible technique to probe the monolayer molecules. However, given the reflection absorbance of $-\log(R/R_0)$ is in the magnitude of 10^{-4}[24] it is often challenging for identification of the molecules let alone quantitative analysis.

Without any additional considerations or further analysis, nonlinear vibrational spectra therefore present useful qualitative information such as the presence of ordered species at the surface, the surface environment as revealed by the resonant frequency (shifts) compared to the bulk phase IR or Raman values,[25,26] heterogeneity of the surface environment from the line widths, and the polarity of the functional group orientation from the direction in which the bands point in the case of interference with significant non virbational resonant modes. However, further evaluation may be performed in order to extract quantitative information, such as a ratio of a mode amplitude in two different polarization schemes to gain information on the molecular orientation.[27-32] Also, it is often necessary to fit the observed bands to a model even if just to extract the resonant frequency, as the peak value may be shifted due to interference. Many schemes have been proposed for such lineshape analysis.[33–39]

Experimentally, nonlinear vibrational spectroscopic technique, much like infrared absorption measurement, can be perform in both reflection as well as transmission geometry. In both cases, a fixed visible wavelength beam is overlapped with a tunable or broadband infrared beam to generate a third field at the sum or difference frequency.[18,33,40–42] The newly generated field can then be interpreted as a function of the IR frequency to reveal the vibrational information. It is to be noted here that although the third field carries vibrational resonance information, yet it is up-converted to the visible wavelength region and the signal is directly proportional to $\chi^{(2)}$. For quantitative analysis, much like its linear optical counterpart, the measurement is perform in multiple polarization combinations. Each of the polarization combination of the $\chi^{(2)}$ spectrum is referred to as $\chi^{(2)}$ effective, denoted as $\chi^{(2)}_{\text{eff}}$, which contains both contributions of one or more of the $\chi^{(2)}_{ijk}$ out of the 27 tensor elements in the laboratory frame in the Cartesian coordinates as well as its respective local field effect correction factor. $\chi^{(2)}_{ijk}$ is then related to the molecular hyperpolarizability which contains molecular orientation information. The orientation analysis can be perform when the ratio of the two unique $\chi^{(2)}_{ijk}$ is obtained and compared with existing model to extract orientation information of the functional groups relative to the laboratory frame.

Additionally, the intensity measurement from a SFG spectroscopy experiment is also known as the homodyne SFG, which it yields the signal that is proportional to the $|\chi^{(2)}_{\text{eff}}|^2$. Phase-sensitive or heterodyne (PS-SFG) SFG yields both the magnitude and phase of the $\chi^{(2)}_{\text{eff}}$. Given that at each frequency, $\chi^{(2)}_{\text{eff}}$ is a complex value, therefore the phase information is loss with the homodyne SFG. Heterodyne SFG, on the other hand, combines SFG with interferometry to afford the complete picture of the second order susceptibility.

1.2.3 SFG studies of organic molecules on metal surfaces

Imagine a system with a monolayer of molecules structured onto a dielectric substrate such as fused silica. For the scanning mid infrared-visible SFG scheme where a tunable mid-IR beam is overlapped with a fixed visible beam to generate the sum frequency field. The scanning mid-IR beam is scanning over the region of the vibrational resonances of the functional groups from the molecules of interest. Therefore, the resulting SFG spectra appears similarly to an IR absorption spectra and the visible beam does not contribute to the overall spectral lineshape. Thus, there is no dispersion and often no SFG signal outside of the region of the vibrational resonant modes of the adsorbed molecules. However, this is not the case for system with most metal substrate where the fixed visible beam can be in resonance with the electronic transition of the metal substrate itself creating a non vibrational resonant but electronically resonant enhancement to the resulting SFG field. Thus, in a system with such description, we can express the overall $\chi^{(2)}$ contribution as:

$$\chi^{(2)} = \chi^{(2)}_{\text{NR}} + \chi^{(2)}_{\text{R}} \tag{1.5}$$

where the subscript of NR refers to the non vibrational resonant in vibrational SFG spectroscopy. Such contribution gives the overall spectral a much complicated appearance given that $\chi^{(2)}$ is a complex value. The cross term of the resonant and non-resonant contribution enhances the resonant features. As well, given the phase of the $\chi^{(2)}_{\text{NR}}$, the resonant modes can appear either a peak or a trough. Such complications add an additional layer of difficulty when analyzing the SFG spectrum quantitatively. Additionally, given the nature of the optical constants with significant adsorption coefficient from the metals, the treatment of the local field effect correction factor compounds the additional contribution to the SFG spectrum.

In the field of sum frequency generation studies of metal interfaces various studies for modelling interfaces closely resemble that of molecular electronic, bioelectronic, photo-voltaic devices and eletrochemical cell have been realized in the past two decades.[32, 43–50]

Although, it is a challenging interface given the numbers of contribution affecting the resulting spectral, many research groups have proposed and demonstrated unique approaches to these obstacles.

In the early work of Davies et al. where dodecanol is adsorbed onto a vapour-deposited gold surface in solution phase, they have utilized P-polarized vibrational SFG spectroscopy to study the polar orientation of dodecanol.[44] In this work, they have experimentally verified that when surfactants adsorb at the interface between water and a hydrophobic surface, the surfactants align themselves in such a way that the tail group is facing the hydrophobic phase while the head group is point towards the solution. The work is done by considering two $\chi^{(2)}$ tensor elements contributing to the $\chi_{\text{eff}}^{(2)}$, namely the zzz and xxz components. By first determine the relative phase of the mode of the methyl group from the dodecanol along with a well established knowledge that long chain alkanethiols (ODT) bind to gold through the sulfur atom and have its methyl group pointing at the environment, they then use this fact to compare the experimentally determined polar orientation of the dodecanol under the same system while swapping out the ODT to its deuterated version. By comparing the phase obtained from both the d-ODT and dodecanol, they have concluded that the terminal methyl group on dodecanol is on average pointing towards the metal surface in aqueous solution phase give that the methyl group has an opposite sign to that of the ODT. As one of the earlier SFG studies on organic-metal interface, the study has served as a good benchmark in the field of SFG for metal interfaces.

Vibrational sum frequency generation spectroscopy has also been employed in situ, in studying eletrochemical processes at the molecular level. In 2014, Baldelli et al. have demonstrated an SFG experiment looking at the reductive desorption of alkanethiols on the working electrode of an eletrochemical cell.[51] The spectra of alkanethiols on gold electrode are acquired at various potentials to monitor the vibrational resonant modes. They have found that negative potentials result in the disordering of the alkyl chains. Furthermore, they have also observed the desorption of the alkanethiols as a result of the applied voltage.

Such experiments have shown the versatility of SFG as a non-destructive spectroscopic approach to afford a molecular understanding of the interface in real time.

In 2019, Tahara et al. published an important demonstration on the advancement of extracting polar orientation from liquid-metal interface with the help of heterodyne sum frequency generation and some clever approach on the reference sample.[52] Such approach is especially useful for metal interfaces given the additional local field effect correction factor being complex values as well as the polarization geometry limitation in most cases. By using HD-SFG, Tahara and co-workers have presented a proof of concept experiment where the polarity of the ODT methyl group is revealed experimentally on platinum surface in the water-platinum interface. By using the knowledge that at air-metal interface, surfactants such as ODT structured in such a way that the terminal methyl group is pointing towards the environment side, they have determine the polar orientation of the terminal methyl on ODT in the liquid phase. Such conclusion is drawn by comparing the sign of the amplitude of the vibration mode of the methyl symmetric stretch bewteen air-metal interface and liquid-metal interface. Furthermore, this work has also provided a novel method of the phase reference that is crucial to the HD-SFG experiment by vapour deposit the metal substrate onto a piece of α-quartz with its phase of $|\chi_{\mathrm{eff}}^{(2)}|^2$ known.

1.3 Scope of the book

As there are several obstacles in the analysis of SFG spectra for organic-metal system especially for substrates that has not been well characterized or studied in the field of nonlinear optics, I first demonstrate the capability of the combination of the homodyne and heterodyne SFG in determining the polar orientation of the function groups on metals. As well, I have proposed a measurement and analysis scheme that is capable of extracting all $\chi_{ijk}^{(2)}$ elements in a p-polarized multi- angle of incidence heterodyne experiment of which the $\chi_{\mathrm{eff}}^{(2)}$ is a linear combinations of $\chi_{ijk}^{(2)}$. Finally, all of these approaches are combined in order to study the structure of a phosephate ester surfactant on iron. This employs a combination of spectroscopic techniques and eletrochemistry in an attempt to understand how the molecular orientation of the surfactant might contribute to its effectiveness as a corrosion inhibitor.

Chapter 2

Background

2.1 Vibrational SFG spectroscopy

2.1.1 SFG intensity and effective susceptibility

The generated sum frequency (SF) beam from a vibraiontal SFG measurement can be described as a plane wave which has a magnitude and a phase that are related to the incident probing beams. By analyzing the SFG spectra taken with different incident beam angles and/or polarizations, the molecular orientation can be deduced.[29,42,53–62]

The second order nonlinear susceptibility, $\chi^{(2)}$, is a third rank tensor describing the relationship between the second order polarization and the two incident fields. In general, $\chi^{(2)}$ can be related to the SFG intensity in reflection geometry by

$$I(\omega_{\text{SFG}}) = \frac{8\pi^3 \omega^2 \sec^2 \theta_{\text{SFG}}}{c^3 n_1(\omega_{\text{SFG}}) n_1(\omega_1) n_1(\omega_2)} |\chi_{\text{eff}}^{(2)}|^2 I(\omega_1) I(\omega_2) \tag{2.1}$$

where ω_{SFG}, ω_1, ω_2 are the frequency of the SFG, incident visible and IR fields, respectively. $n_i(\omega_i)$ is the refractive index of the bulk medium i at the frequency ω_i and θ_{SFG} is the reflected angle of the SFG field from the interface normal.

In addition, the effective second order nonlinear susceptibility can be further expanded in to

$$\chi_{\text{eff}}^{(2)} = L_{ii} L_{jj} L_{kk} e_i e_j e_k \chi_{ijk}^{(2)} \tag{2.2}$$

where the L is a local field correction factor that relates the laser far field to the local field at the point of the generation of the nonlinear field and the e_{ijk} is the respective unit polarization vector relating the fields described in either s- or p- polarization to the laboratory frame in a Cartesian coordinate system.

2.1.2 Local field considerations

Local field correction factor plays an important role in analyzing SFG spectra when the dispersion of the correction factor itself contributes to the $\chi_{\text{eff}}^{(2)}$ lineshape. The local field effect is not a nonlinear optical phenomenon rather it is purely a superposition of the incident field and the reflected field at the point of incidence. The result can be obtained from

$$\begin{aligned}
E_{\text{local}} &= E_{\text{i}} + E_{\text{r}} \\
&= E_{\text{i}} + r \cdot E_{\text{i}} \\
&= (1 + r)E_{\text{i}},
\end{aligned} \tag{2.3}$$

where E_{local} is the local field, E_{i} is the incident field and E_{r} is the reflected field. Thus, the factor of $1 + r$, where r is the Fresnel coefficient of reflection, is known as the local field correction factor, L. By convention, Fresnel coefficient of reflection is represented by the optical polarization of s- and p- which can be described by

$$r_p = \frac{n_2 \cos\theta_1 - n_1 \cos\theta_2}{n_1 \cos\theta_2 + n_2 \cos\theta_1} \tag{2.4a}$$

$$r_s = \frac{n_1 \cos\theta_1 - n_2 \cos\theta_2}{n_1 \cos\theta_1 + n_2 \cos\theta_2} \tag{2.4b}$$

and, for both polarizations $n_1 \sin\theta_1 = n_2 \sin\theta_2$ according to the Snell's law. Therefore, the unit polarization vector is necessary for the transformation to the Cartesian coordinates. If assume that the incident plane is that of the xz-plane where z-axis is the surface normal, then the complete expression of the local field correction factor can be expanded as

$$L_{xx} = \frac{2n_1 \cos\theta_2}{n_1 \cos\theta_2 + n_2 \cos\theta_1} \tag{2.5a}$$

$$L_{yy} = \frac{2n_1 \cos\theta_1}{n_1 \cos\theta_1 + n_2 \cos\theta_2} \tag{2.5b}$$

$$L_{zz} = \frac{2n_1^2 n_2 \cos\theta_1}{n_1 \cos\theta_2 + n_2 \cos\theta_1} \left(\frac{n_1}{n'}\right)^2, \tag{2.5c}$$

where n is the refractive index, θ is the angle of incidence while the subscript 1,2 refers to the two media. The additional factor of $(n_1/n')^2$ in the L_{zz} expression originates from the discontinuity of the normal component (z) of the electric field. Since the normal component of the electric field (with respect to the interfacial plane) is discontinuous across the interface, a phenomenological term, n', is often used in the local field expression. The term n', or the effective refractive index, is derived from a three-layer model of an interface with the middle interfacial layer being an infinitesimally thin sheet sandwiched between two bulk layers. Given that the refractive index is a macroscopic property of a bulk phase, n' cannot be thought of as the refractive index of the interface. Rather, it is the ratio of the two anisotropic microscopic local field factors at the molecular level at the interface in reference to 63. From the expression of the local field correction factor, it is obvious to understand its potential lineshape contribution to the SFG signal given the dispersive nature of the refractive index.

2.1.3 Unit polarization vectors

The Cartesian projection of the s- and p- polarized electric fields is noteworthy to mention here as well as to clarify the convention used for the definition of the coordinate system. The incident plane is assumed to be the xz-plane by convention. The field polarized parallel to the plane of incidence (the plane containing the surface normal z and the incident wavevector) can therefore be split into an x- and z- components obtained using the $e_x = \cos\theta$ and $e_z = \sin\theta$ projections, respectively. Given that these are vector quantities, the x- component of the p-polarized electric field switches sign when describing the reflected field. Therefore, if by assuming that the incident field is the reference positive direction, then the x-component of the SFG field should take on a negative sign. For example, $L_{xx}e_xL_{xx}e_xL_{zz}e_z\chi_{xxz}^{(2)}$ should be written as $-L_{xx}e_xL_{xx}e_xL_{zz}e_z\chi_{xxz}^{(2)}$, where the subscript order is SFG, visible and IR field, respectively.

2.1.4 Vibrational lineshapes

Vibrational sum-frequency generation spectroscopy can provide valuable qualitative and quantitative information about molecular species at surface and buried interfaces. For example, the resonance frequency of a particular chemical function group is revealing of the surface environment, especially when compared to what is observed in bulk IR absorption or Raman scattering spectra. Furthermore, the amplitude of the mode can be related to the molecular orientation, providing a detailed quantitative account of the surface structure. Each of these attributes, however, requires fitting the spectra to some vibrationally resonant lineshape.

The frequency-dependence of the second order nonlinear susceptibility, $\chi^{(2)}$ is the dispersion expression of

$$\chi^{(2)}(\omega_{\mathrm{IR}}) = |\chi^{(2)}|_{\mathrm{NR}} e^{i\phi_{\mathrm{NR}}} + \sum_q |\chi^{(2)}|_{\mathrm{R}} e^{i\phi_{\mathrm{R}}} \tag{2.6}$$

where $\chi^{(2)}_{\mathrm{NR}}$ and $\chi^{(2)}_{\mathrm{R}}$ are the non-vibrational resonant contribution and resonant contribution of the second order nonlinear susceptibility.

The homogeneous component of the lineshape is represented in the frequency domain by a Lorentzian function

$$f_L(\omega) = \frac{A}{\omega_0 - \omega - i\Gamma} \tag{2.7}$$

where A the amplitude, ω_0 is the resonant frequency, and Γ is the homogeneous or intrinsic linewidth, proportional to the reciprocal vibrational dephasing time T_2.[64,65] If multiple such modes are observed, regardless of whether they are well-separated in frequency or overlapping, the frequency dependence of the second-order susceptibility may be described by

$$\chi^{(2)}(\omega) = \chi^{(2)}_{\mathrm{NR}} + \sum_q \chi^{(2)}_q$$

$$= A_{\mathrm{NR}} e^{i\phi_{\mathrm{NR}}} + \sum_q \frac{A_q}{\omega_q - \omega - i\Gamma_q} \tag{2.8}$$

where each of the q modes has a corresponding Lortzentian amplitude A_q, resonance frequency ω_q and width Γ_q, and $\chi_{NR}^{(2)} = A_{NR}e^{i\phi_{NR}}$ represents any vibrationally non-resonant contribution, if present. We note that for dielectric materials, $\chi_{NR}^{(2)}$ is real since $\phi_{NR} = \pm 180°$, but in the case of materials that are not transparent to the visible and SFG beams such as metals, $\chi_{NR}^{(2)}$ may be complex-valued.

In theory, there should also be line broadening due to instrumental considerations (laser and spectrograph bandwidth) and distribution of molecular environments. Such broadening mechanisms are well-described by a Gaussian function

$$f_G(\omega) = A \exp\left[-\frac{(\omega - \omega_0)^2}{2\Gamma_G^2}\right] \tag{2.9}$$

where A is the peak amplitude, ω_0 is the resonance frequency, and Γ_G is the net inhomogeneous linewidth. In practice, the signal-to-noise ratio is often insufficient to distinguish whether a band has a pure Lorentzian profile. In the case of IR absorption and spontaneous Raman scattering spectra, fitting may be performed using a purely Gaussian lineshape, as there are no interference effects. Those bulk spectroscopic techniques directly measure the imaginary component of the first- and third-order susceptibility, respectively. In the case of SFG spectroscopy, however, Gaussians alone do not provide the phase contribution to account for spectral interference. In cases of low signal-to-noise or insufficient spectral resolution in SFG, the default approach is to fit using purely Lorentzian lineshapes, letting Γ be as wide as necessary to fit the bands reasonably well. In such cases, one could report widths as large as 10–15 cm^{-1}, even though previous measurements of the dynamics reveal that the homogeneous linewidth should be much narrower.[33,34,66,67]

A more rigorous approach is to employ a lineshape that uses the known homogenous width Γ_L, and fit to the experimental data to obtain the inhomogeneous component Γ_G. A popular example is the Faddeeva function[68]

$$\begin{aligned}
f_V(\omega) &= f_L(\omega) \otimes f_G(\omega) \\
&= \int_0^{\infty} \frac{A}{\omega_L - \omega - i\Gamma_L} \exp\left[\frac{(\omega_L - \omega_0)^2}{2\Gamma_G^2}\right] d\omega_L
\end{aligned} \tag{2.10}$$

derived from a convolution of Lorentzian and Gaussian lineshapes by introducing an additional inhomogeneous decay term.[69,70] When the spectral resolution and signal-to-noise are sufficiently high, it may be possible to determine both $\Gamma_{L,q}$ and $\Gamma_{G,q}$ from a fit to the data.[71,72] This is especially true in heterodyne experiments, since simultaneous fitting of the real and imaginary $\chi^{(2)}$ spectra facilitates this separation of homogeneous and inhomogeneous components.[73,74]

2.1.4.1 Treatment of local field corrections

Experimental data from a homodyne SFG experiment representing $|\chi^{(2)}|^2$ may be fit to the magnitude squared of Eq. 2.8. Note that if the dispersion of the local field corrections is not significant over the measured spectral range, it is possible to fit $|\chi^{(2)}_{\text{eff}}|^2$ using Eq. 2.8, and then apply local field corrections $\mathbf{L}$ to the extracted A_q.[20,75] For example, in the case of SSP (s-polarized SFG, s-polarized visible, p-polarized infrared) and SPS polarization schemes,

$$\chi^{(2)}_{\text{ssp}} = L_{yy}(\omega_{\text{SFG}})e_y L_{yy}(\omega_{\text{vis}})e_y L_{zz}(\omega_{\text{IR}})e_z \chi^{(2)}_{yyz} \tag{2.11a}$$

$$\chi^{(2)}_{\text{ssp}} = L_{yy}(\omega_{\text{SFG}})e_y L_{zz}(\omega_{\text{vis}})e_z L_{yy}(\omega_{\text{IR}})e_y \chi^{(2)}_{yzy}. \tag{2.11b}$$

Each of these polarizations probes only a single element of $\chi^{(2)}$ in a sample with axial ($C_{\infty v}$) symmetry about the surface normal. As a result, if the peak amplitudes in Eq. 2.8 $A_{q,yyz}$ and $A_{q,yzy}$ are desired for an orientation analysis, there are two options. The first is to extract $\chi^{(2)}_{yyz}$ and $\chi^{(2)}_{yzy}$ from the spectra (or $|\chi^{(2)}_{yyz}|^2$ and $|\chi^{(2)}_{yzy}|^2$ in the case of homodyne data) by calculating $L_{ii}(\omega_{\text{SFG}})L_{jj}(\omega_{\text{vis}})L_{kk}(\omega_{\text{IR}})$ (or the magnitude squared of this quantity for homodyne data), and then correcting the measured intensity. The second option is to fit the intensity data to Eq. 2.8 directly to obtain $A_{q,\text{eff}}$, and then use the appropriate ratio of $L_{ii}L_{jj}L_{kk}$ elements when interpreting the $A_{\text{SSP}}/A_{\text{SPS}}$ ratio to extract the molecular orientation dependence of the underlying A_{yyz}/A_{yzy}. It should be emphasized however that Eq. 2.8 is an expression for $\chi^{(2)}$ itself, and not $\chi^{(2)}_{\text{eff}}$. Any dispersion in $\mathbf{L}$ alters the lineshape, and is therefore best handled prior to fitting. Note that fitting the effective susceptibility is the only option available when treating PPP spectra, since the relative weighting of the four contributing $\chi^{(2)}$ elements (*xxz*, *xzx*, *zxx* and *zzz*) is generally not known in advance.

2.1.4.2 Selection of fitting algorithm

Optimizers can generally be separated into categories depending on whether they require only function evaluations or first derivatives as well (these can always be numerically evaluated using additional function evaluations if analytical forms are not tractable), whether they are bounded (including the option of permitting and/or requiring upper and lower bounds for each parameter), their performance (speed and accuracy), susceptibility to local minima, and sensitivity to initial conditions. When fitting IR or Raman spectra, the amplitudes of all modes are positive, and nearly any algorithm can be used to obtain the same results. It is therefore not typical to discuss the algorithm, although common choices are nonlinear least squares techniques such as Levenberg-Marquardt.[76] When fitting SFG spectra, however, the situation is much less straightforward, and not all algorithms will return the same result. There are two problems: the first is that, on resonance ($\omega_{IR} = \omega_q$), each Lorentzian has a value of A_q/Γ_q. A small change in Γ_q to increase the width of the lineshape will require A_q to be increased accordingly in order to have only a small change in the residual. In other words, depending on the sequence in which the parameters are adjusted, and whether they are varied individually or in pairs/groups, the fitting may converge prematurely. The second and more serious issue is that, if the relative sign of A_q between two peaks is not known (as is generally the case), it is very difficult to "force" an optimizer to explore both options with the same effort, i.e. varying the resonant frequencies, amplitudes, and widths in order to determine the best fit combination with all sign combinations in the numerator of Eq. 2.8. Almost all optimizers struggle with this, and approaches based solely on least squares algorithms are likely to give up early. The most serious consequence is that there is little (and often no) indication of this. The residuals may be small, and the fit may be qualitatively decent in comparison to the experimental data—but the global minimum, or a deeper local minimum, may exist when one of the resonant modes is phase shifted by 180°. This is critical not only for obvious applications such as bond polarity determination, but also for basic analysis such as peak amplitude ratios and resonance frequency determination, as these parameters are all affected by the presence of neighboring vibrational resonances.[77] One approach that has been employed in our laboratory is to first generate all combinations of signs for the number of peaks to be employed in the fitting, then explore them exhaustively for simple cases, or sampled randomly (Monte Carlo techniques can be employed here) when the parameter space is too large.[58] For each combination, a quick optimization can then be employed using a technique such as a truncated Newton's method,[78] followed by a clean-up in the vicinity of the minimum using a steepest descent optimizer like a simplex routine.

It should be emphasized, however, that all of the above approaches rely on a *convenient* amount of interference between vibrational modes. When both the spectral resolution and signal-to-noise are sufficient, it should be possible to extract the relative phase. We are referring to a situation where the spectral overlap between neighboring modes is enough to notice whether the resulting interference pattern originates from modes that are in-phase or out-of-phase, but the resonant frequencies are not so close together to again obscure this information.[62] Fig. 2.1 illustrates the case where two neighboring modes have the same separation in their resonant frequencies, same homogeneous linewidths, same magnitude of their amplitudes, but opposite sign. One can see that while the absolute and relative phase is clearly revealed in the real and imaginary $\chi^{(2)}$ spectra from heterodyne data, the homodyne data shows a difference in shape that requires sufficient resolution and signal quality to differentiate. The challenge is to use the correct number and type of resonant lineshape functions, as the fitting routines at best will simply return the parameters associated with those functions. We note that, in some cases, such challenging situations are greatly assisted by the measurement of spectra in multiple beam polarizations. For example, Zhang *et al.* have illustrated that it is possible to distinguish cyano mode vibrations that are 6 cm^{-1} apart by using all three unique polarization schemes (ssp, sps, and ppp) available to achiral systems with $C_{\infty v}$ symmetry.[79]

2.1.5 Heterodyne SFG measurements

Heterodyne or Phase-sensitive sum frequency generation spectroscopy (PS-SFG) is emerging to be a powerful technique in the SFG community. The main difference between PS-SFG and the conventional homodyne SFG comes down to the detection method:For homodyne SFG the intensity of $\chi^{(2)}$ is collected whereas PS-SFG is capable of probing the magnitude and absolute phase of $\chi^{(2)}$. The additional phase information can help deduce the polarity of the functional group (i.e. whether the functional group is pointed up or down with respect to the substrate). Thus, PS-SFG is an attractive method to distinguish the 180° phase ambiguity from the traditional spectral fitting routine of the homodyne scheme since the absolute phase is determined.[46,81–86] The combination of the additional phase information and $|\chi^{(2)}|$ yields the imaginary $\chi^{(2)}$ spectrum which enables a direct spectral comparison to the bulk response from IR and Raman spectroscopy.

The basis of the phase measurement is to measure the intensity of the signal I when the SFG field from the sample E_{S} and the local oscillator (LO) at the same frequency E_{LO} are brought to coincide. We can write this as

$$
\begin{aligned}
I &= |E_{\mathrm{S}} + E_{\mathrm{LO}}|^2 \\
&= |E_{\mathrm{S}}|^2 + |E_{\mathrm{LO}}|^2 + 2|E_{\mathrm{S}}||E_{\mathrm{LO}}|\cos\Delta\phi
\end{aligned}
\tag{2.12}
$$

where $\Delta\phi$ is the phase difference between sample and local oscillator SFG fields.

As an example for our setup of the phase measurement, a collinear beam geometry is used in the setup and generate the LO in transmission from a piece of y-cut alpha quarts. The three collinearlly travelled fields are then going through a piece of fused silica (phase shifting unit, PSU) of which it rotates about its center from -45˙ to 45˙. By rotating the PSU, the LO is delayed temporally with respect to the sample SFG field thus creating

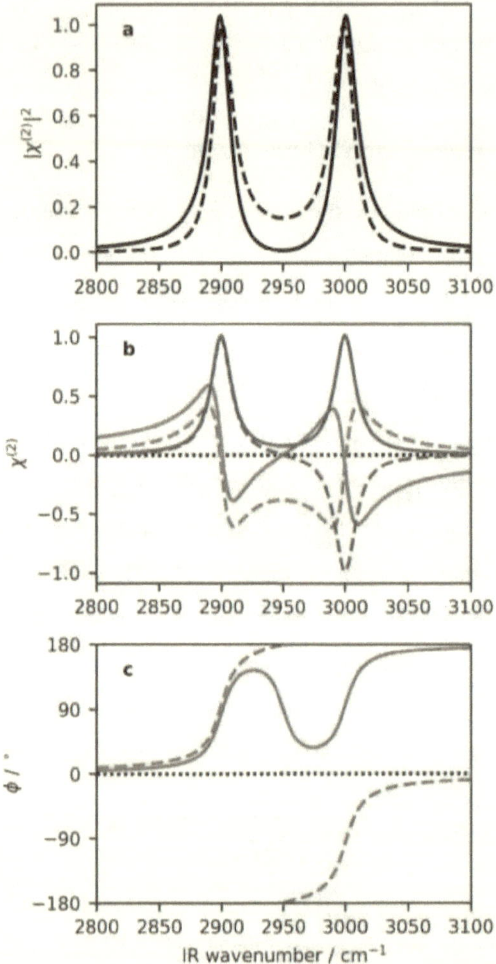

Figure 2.1: Illustration of the effect of relative phase on neighboring vibrational modes. The (a) $|\chi^{(2)}|^2$ intensity, (b) $\mathbb{R}\mathrm{e}\{\chi^{(2)}\}$ and $\mathbb{I}\mathrm{m}\{\chi^{(2)}\}$ spectra in red and blue respectively, and (c) phase for two modes with the same (solid lines) and opposite signs, i.e. phase shifted by 180° (dashed lines) on resonance with respect to each other. One notices that while $\mathbb{R}\mathrm{e}\{\chi^{(2)}\}$ and $\mathbb{I}\mathrm{m}\{\chi^{(2)}\}$ immediately display the relative (and absolute) phase, the relative phase information is subtle in the $|\chi^{(2)}|^2$ spectra. Reprinted with permission from Ref. 80. Copyright 2018 American Institute of Physics.

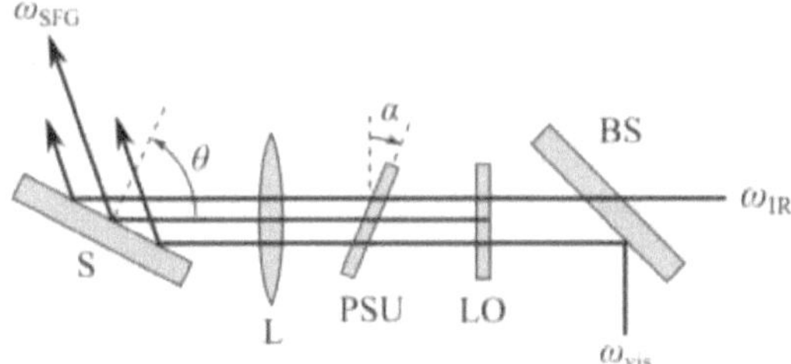

Figure 2.2: Experimental collinear phase measurement scheme. Reprinted with permission from Ref. 87. Copyright 2015 American Chemical Society.

an interferogram as a function of the PSU tilt angle,α. Fig. 2.2 shows the experimental scheme.

The interferogram obtained at each frequency as a function of the the PSU tilt angle can be expressed as:

$$I(\alpha,\omega) = a(\alpha,\omega) + b(\alpha,\omega)\cos\Delta\phi(\alpha,\omega) \tag{2.13}$$

where $a \propto |E_S|^2 + |E_{LO}|^2$, $b \propto 2|E_S||E_{LO}|$, and ω refers to the frequencies of all three beams. In addition, the experimentally observed phase term, Δ_ϕ in Eq. 2.13 can be further separated into individual components related to all, PSU, the focusing lens, as well as the phase difference between the sample SFG field and the LO as described as

$$\Delta\phi(\alpha,\omega) = \Delta\phi_{PSU}(\alpha,\omega) + \Delta\phi_{lens}(\omega) + \Delta\phi_{S-LO}(\omega). \tag{2.14}$$

The description of the phase shift regarding with the PSU is a result of the differing optical path lengths through the fused silica as shown in Fig. 2.3. Although the incident angle of the three beams are the same in the collinear geometry scheme. However, the refracted angles, β_i are different due to the dispersion of the refractive index,

$$\beta_i(\alpha,\omega_i) = \arcsin\frac{n_{1,i}}{n_{2,i}}\sin\alpha, \tag{2.15}$$

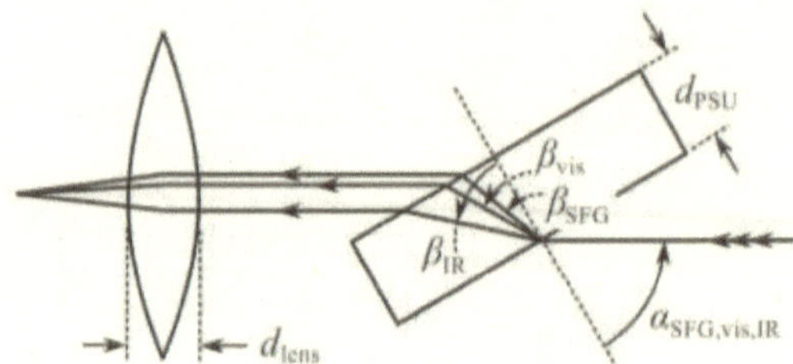

Figure 2.3: Scheme of the phase shifting unit in top-down view. Reprinted with permission from Ref. 87. Copyright 2015 American Chemical Society.

where i refers to each of the three beams, and n_1 and n_2 refer to the respective frequency dependent refractive indices of the incident medium and the refracted medium. The phase difference then comes from the delay between the LO and the two input beams passing through the PSU which can be described as

$$\Delta\phi_{PSU}(\alpha,\omega) = \frac{d_{PSU}}{c}\left(n_{SFG}\,\omega_{SFG}\cos\beta_{SFG} - n_{vis}\,\omega_{vis}\cos\beta_{vis} - n_{IR}\,\omega_{IR}\cos\beta_{IR}\right), \qquad (2.16)$$

where d_{PSU} is the thickness of the PSU, and c is the speed of light.

The phase shift associated with the focusing lens in Eq. 2.14 can be determined in a similar fashion as

$$\Delta\phi_{lens}(\alpha,\omega) = \frac{d_{lens}}{c}\left(n_{SFG}\,\omega_{SFG} - n_{vis}\,\omega_{vis} - n_{IR}\,\omega_{IR}\right). \qquad (2.17)$$

Upon fitting the interferogram as shown in Fig. 2.4 as an example, while considering all three phase terms from Eq. 2.14, $\Delta\phi_{S-LO}$ can be determined. In order to measure the phase of an unknown sample, $\phi_{S.LO}$ needs to be interpreted with respect to a known standard. Popular choices include molecular films with known polarity due to chemical immobilization such as OTS (octadecyltrichlorosilane), methoxy, or alkane thiol/gold; molecular films with known polarity due to hydrophobic interactions such as PDA, long-chain alcohols (dodecanol) at the air–water interface; or crystals with bulk nonlinear

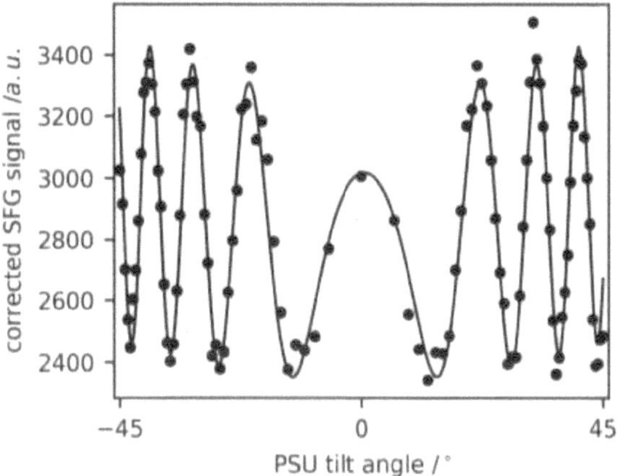

Figure 2.4: Example of the temporal interferogram from the heterodyne phase measurement of the z-cut α-quartz at 2800 cm^{-1}.

susceptibilities such as α-quartz. Note that in order to use α-quartz as a standard, it must be known whether the material is left- or right-handed, which crystallographic face is exposed, and the azimuthal orientation of the exposed face. For example, even in the case of z-cut right-handed α-quartz, it is known that rotation about the z axis changes the sign of the bulk nonlinear susceptibility. As a result of the D_{3h} symmetry about this axis, there will be 6 lobes in $|\chi^{(2)}|^2$, with the phase alternating between $0°$ and $180°$ between each lobe. One approach to resolving the polarity of the $\chi^{(2)}$ phase for a particular quartz sample is to perform piezoelectric measurements. Our approach is by calibrating the phase of a 6-mm thick z-cut sample with etch-marked orientation by comparison to OTS and then gold.

Experimentally, the phase of the $\chi^{(2)}_{\mathrm{eff}}$ can be determined by first obtaining the phase difference of the $\Delta\phi$ between the unknown sample and the substrate with a known phase (in our case, z-cut alpha quartz). The difference in $\Delta\phi$, or $\Delta\Delta\phi$ is the phase difference between the sample and the z-cut alpha quartz. Finally, by adding $\Delta\Delta\phi$ to the phase of the z-cut quartz, we can obtain the phase of the $\chi^{(2)}_{\mathrm{eff}}$ of the unknown sample. It is to be noted that, the phase difference of the the Fresnel coefficient of the reflection between the unknown sample and the phase reference sample also needs to be consider. In an example of a p-polarized heterodyne experiment, the expression can be described as

$$\phi_{\chi^{(2)}_{\text{eff,sample}}} = \Delta(\Delta\phi) + \phi_{\chi^{(2)}_{\text{eff,zcut}}} + \Delta\phi_{r_p,\text{sample}-\text{zcut}}. \tag{2.18}$$

2.1.6 Other methods of obtaining complex-valued $\chi^{(2)}$ elements[*]

2.1.6.1 Kramers-Kronig approaches

If only intensity data is available, there are techniques available to extract the phase that do not assume any particular model for the lineshape, and therefore do not require prior knowledge of the nature or number of resonant modes. One such option uses the causality relationship in the dispersion of the real and imaginary components of $\chi^{(2)}$ in the form of a Kramers-Kronig transformation[88–94]

$$\text{Re}\{\chi^{(2)}\}(\omega) = \frac{1}{\pi} P \int_{-\infty}^{\infty} \frac{\text{Im}\{\chi^{(2)}\}(\omega')}{\omega' - \omega} d\omega' \tag{2.19a}$$

$$\text{Im}\{\chi^{(2)}\}(\omega) = -\frac{1}{\pi} P \int_{-\infty}^{\infty} \frac{\text{Re}\{\chi^{(2)}\}(\omega')}{\omega' - \omega} d\omega' \tag{2.19b}$$

where P is the Cauchy principal value and ω' is a dummy spectral variable over which the integration is performed. The quality of the approximation may be assessed by computing $|\chi^{(2)}|^2 = \text{Re}\{\chi^{(2)}\}^2 + \text{Im}\{\chi^{(2)}\}^2$ and minimizing the residuals compared to the experimental intensity data. There are two caveats associated with this approach. The first is that the integration must be performed over a finite frequency domain in practice, resulting in an unknown offset to $\text{Re}\{\chi^{(2)}\}(\omega)$. For systems with no significant non-resonant contribution, this offset is readily evaluated. The more serious difficulty is that, as only *relative* phase information is encoded in the homodyne data, there is an overall phase offset

$$\chi^{(2)} = |\chi^{(2)}| \exp[i(\phi + \phi_{\text{offset}})] \tag{2.20}$$

---

that affects the real and imaginary components. If one is willing to assume that ϕ_{offset} is independent of ω in the frequency region of interest, then some additional information can be used to resolve this ambiguity. For example, perhaps the phase of the non-resonant response is known, or the polarity (and hence phase) of a well-separated resonant mode is known.

2.1.6.2 Maximum entropy methods

Another option for phase retrieval is to use a concept from information theory, the maximum entropy method.[37, 95–99] This technique seeks to add features to the time domain response function $R^{(2)}(t)$ that is related to the second-order susceptibility through a Fourier transform

$$R^{(2)}(t) = \int_0^\infty \chi^{(2)}(\omega)e^{-i\omega t}d\omega. \tag{2.21}$$

The objective is to add features without increasing the spectral entropy h defined by

$$h = \int_0^1 \log[\chi^{(2)}(v)]dv, \tag{2.22}$$

where

$$v = \frac{\omega - \omega_1}{\omega_2 - \omega_1}$$

is defined such that $0 \leq v \leq 1$ in the frequency region between $\omega_1 = 2800 \text{ cm}^{-1}$ and $\omega_2 = 3800 \text{ cm}^{-1}$ in our example. In the first step, a $|\chi^{(2)}|$ spectrum of $2N+1$ frequencies is Fourier transformed to obtain $R^{(2)}(t)$. One then solves the matrix equation

$$\begin{bmatrix} R_0^{(2)} & R_1^{(2)*} & \cdots & R_N^{(2)*} \\ R_1^{(2)} & R_0^{(2)} & \cdots & R_{N-1}^{(2)*} \\ \vdots & \vdots & & \vdots \\ R_N^{(2)} & R_{N-1}^{(2)} & \cdots & R_0^{(2)} \end{bmatrix} \begin{bmatrix} 1 \\ a_1 \\ a_2 \\ \vdots \\ a_N \end{bmatrix} = \begin{bmatrix} b \\ 0 \\ 0 \\ \vdots \\ 0 \end{bmatrix} \tag{2.23}$$

where N values of a and a single value b are sought. The complex spectrum is finally generated from

$$\chi^{(2)}(v) = \frac{be^{i\phi_{\text{offset}}}}{1 + \sum_{n=1}^{N} a_n e^{inv}}. \tag{2.24}$$

Just as in the case of Kramers-Kronig transformations, It is to be noted that there is a phase offset ϕ_{offset} that is left unresolved. This is to be expected for any method that takes only the magnitude $|\chi^{(2)}|$ into consideration, as any $\phi_{\text{offset}}(\omega)$ will reproduce the magnitude or intensity spectrum. A detailed discussion about the application of MEM to phase retrieval in SFG spectroscopy, including the inner workings of the technique is found in Ref. 95. A discussion of cases where MEM is most successful in SFG spectroscopy, and a consideration of its limitations appears in Ref. 97.

2.2 Spectroscopic Ellipsometry

In order to to analyze the SFG spectra quantitatively, accurate optical constants are necessary in order to obtain the correct molecular orientational information. For obtaining experimental refractive index, ellipsometry is a well suited indirect technique of which the optical constants can be calculated. The basic idea ellipsometry is the measurement of the change in polarization of the reflected light. The ratio of the two optical polarization light, r_p/r_s or ρ can be obtained from the detector signal using

$$\frac{r_p}{r_s} = \frac{E_p^{out}/E_p^{in}}{E_s^{out}/E_s^{in}} = \frac{|E_p^{out}|/|E_p^{in}|}{|E_s^{out}|/|E_s^{in}|} e^{i(\delta_p - \delta_s)} = \tan\Psi e^{i\Delta} = \rho, \tag{2.25}$$

where the $\tan\Psi$ term is the magnitude of the ratio of the two reflection coefficients and the Δ is its phase. For the calculation the optical constants from experimentally determined ρ, some multiple beam interference model is required for multi-layer samples. However, for interface that is comprise of two semi-infinite thick layers (e.g. air-metal), optical constants can be directly calculated by expending out the Fresnel's reflection of coefficients as described by

$$\rho = \frac{r_p}{r_s} \tag{2.26}$$

using the definitions of r_p and r_s provided in Eqs. 2.4. As we substitute $n_1 = 1$ for air, optical constants can then be directly calculated from

$$\varepsilon(\omega) = n(\omega)^2 = \sin\theta_i^2 \left[1 + \tan\theta_i^2 \left(\frac{1 - \rho(\omega)}{1 + \rho(\omega)} \right)^2 \right], \tag{2.27}$$

where $\varepsilon(\omega)$ is the frequency dependent dielectric constant, $n(\omega)$ is the frequency dependent complex refractive index and θ_i is the angle of incidence of the measurement.[100] Although, the calculation of optical constants can be directly obtained from the single interface inversion equation as described above, it is to be noted here that both Ψ and Δ are dependent on numerous factors such as the film thickness, refractive index, surface roughness, interfacial mixing, composition, crystallinity, anisotropy and uniformity of the substrate.[100] It is for that reason, that elliposmetry is known as an indirectly method of obtaining refractive index. Given a more complex substrate with multiple interfaces and compositions, some model has to be assumed in order to extract useful information from the spectra of Ψ and Δ. If the single interface inversion equation is used to calculate the refractive index of a semi-infinite layer interface, some assumptions have to be taken into consideration as well such as its surface roughness and the uniformity of its composition.

Chapter 3

Determining the Orientation of Chemical Functional Groups on Metal Surfaces by a Combination of Homodyne and Heterodyne Nonlinear Vibrational Spectroscopy[*,†]

3.1 Introduction

Characterizing the structure of molecules attached to metal or semiconductor surfaces is a key step towards the understanding, optimization, and design of catalytic systems, solar cells, and corrosion inhibitors.[9, 101–105] The manner in which molecules adsorb on the metal surface (the nature of their interaction to the surface, their orientation and polarity in the adsorbed state) determine the desired substrate functionality. In industrial applications such as corrosion inhibition, surfactant coatings are a critical frontline prevention for separating bare metal surfaces from oxygenated species. Such knowledge also provides mechanistic details to bottom-up manufacturing processes such as atomic layer deposition. Of all the techniques that are capable of characterizing molecules on surfaces, ones based on vibrational spectroscopy offer structural sensitivity and the ability to potentially

---

[*]Reproduced in part from Yang, W.-C.; Hore, D.K. "Determining Nonlinear Optical Coefficients of Metals by Multiple Angle of Incidence Heterodyne-Detected Sum-Frequency Spectroscopy." *J. Phys. Chem. C* **121**, 28043 (2017). Copyright 2017 American Chemical Society.

[†]SDS-Al data collection, analysis and model development by Wei-Chen Yang. OTS-glass and ODT-gold data collected by Paul Covert. Electronic structure calculations performed by Dennis Hore.

recognize species based on the vibrational signature. However, conventional IR reflection absorption spectroscopy has limited, and often insufficient, sensitivity for low surface coverage. Nonlinear techniques such as visible-infrared sum-frequency generation (SFG) can offer the benefits of a vibrational optical probe, along with sensitivity to sub-monolayer concentrations.

As a direct consequence of the requirement of non-centrosymmetry to produce SFG signal, the phase of the emitted SFG field carries information on the polar orientation of surface chemical functional groups. By characterizing the phase in addition to the amplitude of the reflected SFG field, it is therefore possible to distinguish whether surface methyl groups, for example, are pointed towards the metal or towards the ambient air. Although several methods have been proposed for phase measurement in heterodyne SFG schemes,[46,81–86,106–115] such information does not strictly require explicit phase measurement. This is because metals often have large electronic (vibrationally non-resonant, NR) contributions to the SFG signal that constructively or destructively interfere with the SFG generated from the organic functional groups on IR resonance.[44,116] This interaction is much the same as achieved with an external phase reference (the local oscillator, LO) in a heterodyne experiment. Several studies have made use of this in the analysis of SFG data.[77,102,117–120] Explicit measurement of the phase has several advantages over reliance on implicit phase interpretation, primarily the result of being able to independently characterize, control, and modulate the LO magnitude and phase. In this work we illustrate that, while heterodyne SFG measurements excel at measuring the phase of bare metals, and of organic layers adsorbed onto dielectric substrates with no appreciable NR response, there are challenges associated in phase characterization of organics on metals. We illustrate this with data from a bare aluminum surface, an organic alkyl surfactant on aluminum, alkyl functionalized glass, and an alkyl thiol on gold. After discussing the interplay between resonant and non-resonant responses for each surface, we present a straightforward method for determining the polarity of surface chemical functional groups using a combination of the heterodyne and homodyne SFG data.

3.2   Methods

3.2.1   Sample preparation

Aluminum coupons 1 mm thick were cut to 25 mm × 25 mm squares and both surfaces were rough finished using a milling machine. The samples were then polished by 320, 600, 1200 and 2000 grit papers in succession, each for 10 min. The polishing step was completed by using 3 μm diamond polishing suspension (Buehler MetaDi Supreme) for 15 min, followed by 0.05 μm alumina (Buehler Masterprep) for 15 min. Polished metal samples were washed with soap and rinsed in 18 MΩ·cm deionized water (Nanopure, Barnstead Thermo) for 10 min, and dried under nitrogen gas. The samples were then successively sonicated in acetone, ethanol, a second acetone step, and finally methanol, each for 30 min. The final treatment was exposure to oxygen plasma for 60 min. SFG spectra of these mirror finish Al pieces with strong specular reflection produced only non-vibrationally resonant signal. Sodium dodecyl sulfate 98% was purchased from Sigma-Aldrich and used without further purification. 0.014 M SDS solution was prepared in 18.2 MΩ·cm. The aluminum coupon was immersed into the solution for two hours then dried under nitrogen.

SFG data for tricholoro(octadecyl)silane monolayers on glass was obtained from a previous study;[46] those samples were prepared according to published methods.[121] The reagent (Aldrich, greater than 90% purity) was used without further treatment in a 4:1 solution of hexadecane and CCl_4. Borosilicate glass microscope slides were cleaned in 110°C piranha for 1 h, rinsed with 18.2 MΩ·cm water, and dried under nitrogen. After immersing the clean substrates, unreacted OTS was removed by rinsing with chloroform, acetone, methanol, and water. The final step was drying at 80°C for 3 h. Similarly, SFG data for octadecane thiol (ODT) monolayers on gold was taken from Ref. 87. Those samples were prepared according to procedures described in an earlier report.[122] Substrates with 100 nm Au deposited on a 5 nm Cr adhesion layer (EMF, Ithaca, NY) were cleaned by sonication in acetone and ethanol, then immersed in 1×10^{-3} M solution of ODT in ethanol for 12 h. Any residual ODT was removed by soaking in fresh ethanol. Samples were subsequently dried under nitrogen.

3.2.2 Homodyne and heterodyne SFG spectroscopy

Our wavelength-scanning SFG system and its configuration for phase measurements has been described in Refs.85 and 46. The essential details are that collinear 20 ps s-polarized visible (100 μJ/pulse) and p-polarized infrared (200 μJ/pulse) beams are incident at $70°$. The local oscillator (LO) is generated in transmission before the sample in a 50 μm piece of y-cut quartz, oriented so that its optical axes are rotated only a few degrees from the plane of the incident beam polarizations. This simultaneously controls (reduces) the amount of LO generated to ensure sufficient contrast in the interference fringes,[85] and ensures that the polarization of the transmitted visible and infrared beams are not appreciably altered as a result of the quartz birefringence. Heterodyne data is collected by sequentially scanning the infrared frequency, and rotating a 1 mm fused silica plate that acts as a phase-shifting unit (PSU) between the sample and the y-cut quartz. These signals display temporal interference along the PSU rotation axis, and spectral interference along the IR frequency axis. After each experiment, the sample was replaced with a reference sample, a piece of z-cut quartz whose phase is known as a result of prior calibration.[46, 59] In general, the bulk $\chi^{(2)}$ tensors of non-centrosymmetric crystals are real far from resonance, and have surface $\chi^{(2)}$ values that are shifted by $90°$ from the bulk. For our z-cut sample, we have marked the orientation of the crystal to produce $\phi_{NR} = -90°$. We have previously described the manner in which the sample and reference heterodyne data may be used together to arrive at the phase of the sample second-order susceptibility.[46, 85]

3.2.3 Electronic structure calculations

We have considered a methyl group in three different chemical environments, next to an OH (methanol), in an ester group, and at the end of an alkyl chain (both ends of methyl hexanoate). Geometry optimization and subsequent Hessian calculation were performed in GAMESS using B3LYP/6-31G(d,p) and a PCM effective solvent model. Dipole moment and polarizability derivatives were then obtained using an explicit finite difference approach whereby the eigenvectors of the Hessian were used to construct seven input geometries that step along the methyl symmetric stretching normal mode. Full details of this procedure for estimating molecular hyperpolarizability tensor elements are given in Ref.58.

3.3   Results & Discussion

3.3.1   Phase contrast in a heterodyne SFG experiment

We will show that the presence of a non-resonant contribution to $\chi^{(2)}$ that is significant compared to the resonant contribution effectively diminishes the phase contrast—the variation in phase upon passing through a vibrational resonance—as would be measured in a heterodyne SFG experiment. In the general case we have

$$\chi^{(2)} = \chi_{NR}^{(2)} + \chi_R^{(2)} = |\chi_{NR}^{(2)} + \chi_R^{(2)}|e^{i\phi}. \tag{3.1}$$

We illustrate the addition of $\chi_{NR}^{(2)}$ and $\chi_R^{(2)}$ in the complex plane in Fig. 3.1a. The graphic depicts the case where $|\chi_{NR}^{(2)}| > |\chi_R^{(2)}|$, and the dashed arrows indicate the phase trajectory of $\chi_R^{(2)}$ as it passes from $0°$ to $180°$. At each value of ω_{IR}, the phase of the overall response in Eq. 3.1 is given by

$$\phi = \arctan\left[\frac{|\chi_{NR}^{(2)}|\sin\phi_{NR} + |\chi_R^{(2)}|\sin\phi_R}{|\chi_{NR}^{(2)}|\cos\phi_{NR} + |\chi_R^{(2)}|\cos\phi_R}\right]$$
$$= \arctan\left[\frac{R\sin\phi_{NR} + \sin\phi_R}{R\cos\phi_{NR} + \cos\phi_R}\right] \tag{3.2}$$

where $R \equiv |\chi_{NR}^{(2)}|/|\chi_R^{(2)}|$. The corresponding experimental observation is that there may be only a small change in ϕ upon passing through the vibrational resonance, compared to the $180°$ change when $\chi_{NR}^{(2)} = 0$. A graphical explanation for this may be seen in Fig. 3.1a. In order to observe the phase change in tuning ω_{IR} through this vibration, there must be a significant corresponding change in ϕ. We formally define the *phase contrast* as the

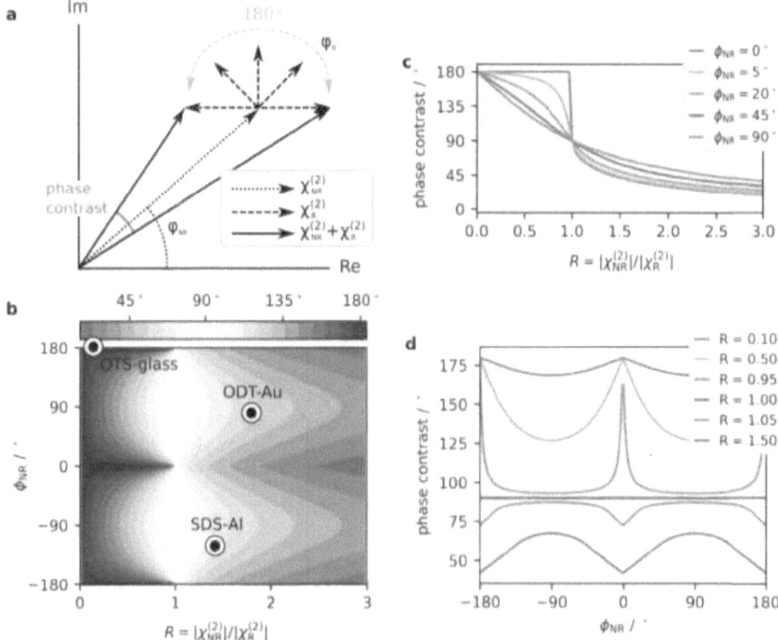

Figure 3.1: (a) Argand diagram illustrating the determination of the phase contrast, and (b) its predicted value according to the ratio of the non-resonant-to-resonant amplitude ratio and the non-resonant phase, ϕ_{NR}. (c) Some slices along the ϕ_{NR} direction and (d) $R = |\chi^{(2)}_{NR}|/|\chi^{(2)}_{R}|$ direction. Reprinted with permission from Ref. 123. Copyright 2017 American Chemical Society.

largest difference in ϕ anywhere in the range $\phi_R = 0$–$180°$ as the resonant mode undergoes a frequency-dependent phase change in the transition from $\omega_{IR} < \omega_0$ to $\omega_{IR} > \omega_0$.

$$\text{phase contrast} = \phi_{\max}(\phi_{NR}, R; \phi_R) - \phi_{\min}(\phi_{NR}, R; \phi_R) \tag{3.3}$$

Note that the best contrast is not necessarily measured pre- and post-resonance, nor does it occur pre- vs on-resonance. Rather, the phase contrast is a function of the ratio R and the value of the non-resonant phase ϕ_{NR}. This is shown in Fig. 3.1b, with some slices along R in Fig. 3.1c, and slices along ϕ_{NR} in Fig. 3.1d.

It is worthwhile to examine some limiting behaviors. When $|\chi^{(2)}_{NR}| \gg |\chi^{(2)}_{R}|$, that is when

$R \to \infty$ the phase contrast approaches zero. This makes sense, as there is no variation in the phase in the absence of any molecular vibrations (or for modes that are not oriented in a polar manner at the surface), and $\phi \to \phi_{NR}$. In the other extreme, as the non-resonant amplitude becomes very small, $R \to 0$ and the phase contrast reaches its maximum value of $180°$, irrespective of ϕ_R. The intermediate cases are of interest, as the contrast depends on both R and ϕ_{NR}. It is interesting to note that when $|\chi_{NR}^{(2)}| = |\chi_R^{(2)}|$ (when $R = 1$), the phase contrast is exactly $90°$, irrespective of ϕ_{NR} (red curve in Fig. 3.1d). However, in the special case where $\phi_{NR} = 90°$, the phase contrast varies sharply for $R < 1$ and $R > 1$, as seen most clearly by the blue curve in Fig. 3.1c. In the case where $\phi_{NR} = 0°$ or $180°$, the phase contrast always displays its maximum value of $180°$ if $R < 1$. We will now compare these predictions with some experimental observations.

3.3.2 Bare metal surface

We first consider a heterodyne measurement of the aluminum surface exposed to air, where the local oscillator is generated in transmission from y-cut quartz. There are no vibrational resonances in the region 2800–3000 cm^{-1} (homodyne spectrum in Fig. 3.2a), but we do observe the interference between the LO and the SFG generated from the Al surface. Data obtained from an experiment where the IR beam frequency is scanned from 2800–3000 cm^{-1}, and the phase-shifting unit is rotated by $90°$ from -45 to $+45°$ is used to obtain the phase information shown in Fig. 3.2b. The interference pattern appears much like one obtained for a transparent bulk nonlinear crystal such as z-cut quartz.[85] However, as the LO reflects off a metallic sample, and subsequently from a dielectric reference sample (z-cut quartz), an additional phase correction is required, as has been described in detail previously.[87] This has already been taken into account in presenting the phase data in Fig. 3.2b, using the frequency-dependent refractive index of aluminum. Using this analysis, we have determined that the phase of $\chi_{NR}^{(2)}$ for our aluminum sample is $-120°$ throughout this region of the mid-infrared. This value depends on the visible wavelength and to the extent that the surface is clean. As the phase of this surface is not a multiple of $90°$, the non-resonant component appears in both $\mathbb{Re}\{\chi^{(2)}\}$ and $\mathbb{Im}\{\chi^{(2)}\}$ as shown in the red and blue lines in Fig. 3.2c.

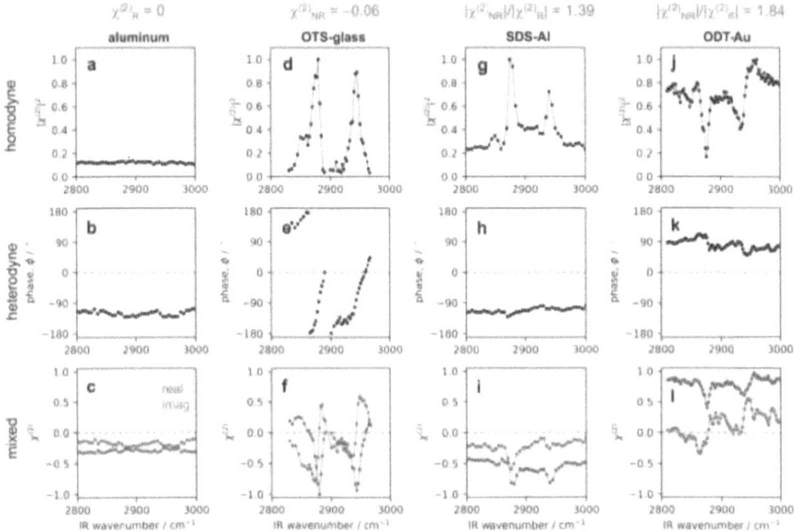

Figure 3.2: The top row shows homodyne SFG data obtained for the bare aluminum surface (left column), glass surface functionalized with OTS (second column), aluminum surface with adsorbed SDS (third column), and gold surface with covalently attached ODT (right column). The middle row shows the phase extracted from a heterodyne SFG experiment from the same four surfaces. The bottom row illustrates the real and imaginary components obtained from $|\chi^{(2)}|\cos\phi$ and $|\chi^{(2)}|\sin\phi$. Reprinted with permission from Ref. 123. Copyright 2017 American Chemical Society.

3.3.3 Glass-organic interface

For comparison, we present another simple case, trichloro(octadecyl)silane (OTS), functionalized onto a glass surface. Homodyne data appears in Fig. 3.2d. As there is negligible non-resonant SFG contribution, so the measured interference in a heterodyne experiment originates primarily between the vibrationally-resonant sample SFG and LO to yield the phase shown in Fig. 3.2e. The real and imaginary components of $\chi^{(2)}$ are indicated by the red and blue traces in Fig. 3.2f. If we fit this data assuming a model where each vibrational mode is represented by a Lorentzian line shape

$$\chi_R^{(2)}(\omega_{IR}) = \frac{A}{\omega_0 - \omega_{IR} - i\Gamma} \tag{3.4}$$

where A is the (signed) amplitude of response, ω_0 is its resonant frequency, and Γ is the homogeneous linewidth, we can estimate the oscillator strength from the expression on resonance as A/Γ. For the methyl symmetric stretch, this produces a value of $R = -0.06$ as indicated by the annotation on Fig. 3.1b, with $\phi_{NR} = \pm 180°$. We therefore predict a phase contrast of nearly $180°$ on passing through this mode near 2875 cm^{-1}, in agreement with our observation.

3.3.4 Metal-organic interface

We now prepare a surface where sodium dodecyl sulfate (SDS) is adsorbed onto the same Al substrate we have previously characterized, and perform a homodyne SFG experiment. The measured signal is a superposition of the Al non-resonant and surfactant resonant contribution as described in Eq. 3.1, and the measured intensity is given by

$$\begin{aligned}
I &\propto \left| |\chi_{NR}^{(2)}| e^{i\phi_{NR}} + |\chi_R^{(2)}| e^{i\phi_R} \right|^2 \\
&= |\chi_{NR}^{(2)}|^2 + |\chi_R^{(2)}|^2 + 2|\chi_{NR}^{(2)}||\chi_R^{(2)}| \cos(\phi_{NR} - \phi_R).
\end{aligned} \tag{3.5}$$

As the IR probe passes through a vibrational mode, we see a frequency dependence in the resonant contribution as in Eq. 3.4. The phase of the resonant component is given by

$$\phi_R(\omega_{IR}) = \arctan\left[\frac{A\Gamma}{A(\omega_0 - \omega_{IR})}\right].$$

Although it appears that the above expression may be independent of A, we have included it in the numerator and denominator, as the sign of A determines the quadrant of ϕ_R, a critical aspect of the phase characterization. In modelling such a response, it is therefore important to preserve the quadrant information in the inverse tangent operation. More explicitly,

$$\phi_R(\omega_{IR}) = \begin{cases} \arctan[\Gamma/(\omega_0 - \omega_{IR})] & \text{for } A > 0 \text{ and } \omega_{IR} < \omega_0 \\ \arctan[\Gamma/(\omega_0 - \omega_{IR})] + 180° & \text{for } A < 0 \text{ and } \omega_{IR} > \omega_0 \\ \arctan[\Gamma/(\omega_0 - \omega_{IR})] - 180° & \text{for } A < 0 \text{ and } \omega_{IR} < \omega_0. \end{cases} \tag{3.6}$$

An important conclusion here is that, when $\chi_{NR}^{(2)} = 0$, ϕ_R changes by $180°$ for an isolated vibrational mode upon passing through resonance. Examining the data Fig. 3.2g relatively far from any vibrational resonance (for example near 2800 cm^{-1}), $\chi_R^{(2)} \approx 0$ and $|\chi_{NR}^{(2)}| \approx \sqrt{I}$ from Eq. 3.5. Estimation of $|\chi_R^{(2)}|$ is not as straightforward as a result of the interference that is described by Eq. 3.5. However, from fitting the data to a sum of Lorentzians plus a non-resonant component, we have determined that $R = 1.39$. We then introduce the local oscillator to perform a heterodyne measurement. The extracted phase is shown in Fig. 3.2h, which bears a striking resemblance to that obtained for the bare Al surface (Fig. 3.2b), with a greatly diminished phase contrast largely blurring the strong vibrational features present in the homodyne data. If we were to combine this phase information with $|\chi^{(2)}|$ obtained from the homodyne data in Fig. 3.2g, we can plot $\chi^{(2)}\cos\phi$ (red trace in Fig. 3.2i) and $\chi^{(2)}\sin\phi$ (blue trace) to obtain the real and imaginary components of $\chi^{(2)}$. In the case of dielectric substrates with negligible (or at least real-valued) $\chi_{NR}^{(2)}$, the sign of $\mathbb{Im}\{\chi^{(2)}\}$ reveals the polarity of the functional group, as will be described in more detail in the following section. Here a complex-valued non-resonant response contributes an offset to both the real and imaginary spectra, so it is now the direction of the band in $\mathbb{Im}\{\chi^{(2)}\}$, and not its sign, that is important.

Before generalizing our approach, we consider one final example of octadecylthiol (ODT) on gold, with the homodyne data in Fig. 3.2j. Gold has been widely used in SFG experiments as a phase reference, including applications where the metal is not in direct contact with the molecules of interest, but close enough to provide a source of non-resonant SFG.[44, 115, 116, 124, 125] For typical beam angles in a reflection experiment, the $|\chi_{NR}^{(2)}|$ for Au is at least two orders of magnitude weaker in ssp than in ppp polarization at 532 nm, often allowing $|\chi_{NR}^{(2)}|$ and $|\chi_R^{(2)}|$ to be comparable.[87] However, Fig. 3.2k again shows that the variation in phase is small. Fitting the data provides $R = 1.84$ for the methyl symmetric stretch, along with our direct measurement of $\phi = 84°$ from the heterodyne experiment. When this point is plotted on the map in Fig. 3.1b, one predicts a phase contrast of roughly

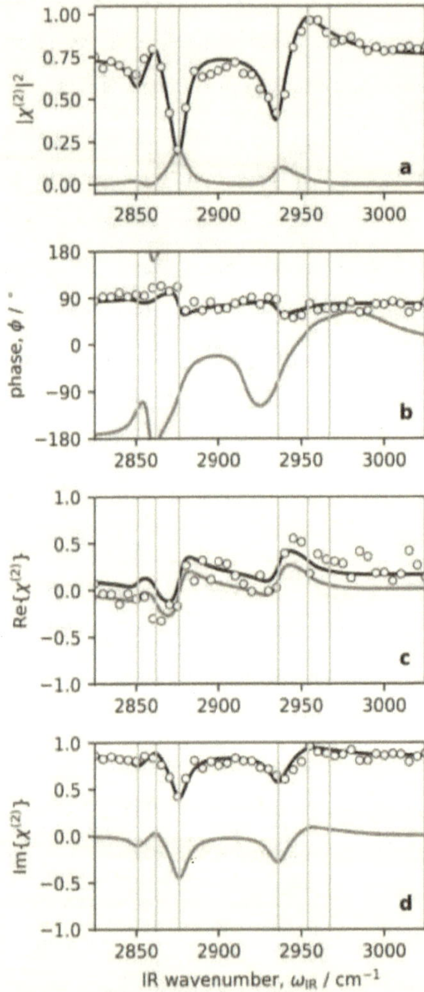

Figure 3.3: Experimental data in points obtained from a determination of the (a) magnitude squared and (b) phase of ODT on gold. These have been transformed into the (c) real and (d) imaginary components of $\chi^{(2)}$ (data in points). A fit to a model where each vibrational mode is represented by a Lorenztian is shown with black lines. The same model, but excluding the gold non-resonant contribution, is plotted with red lines. Reprinted with permission from Ref. 123. Copyright 2017 American Chemical Society.

$60°$, close to what is observed in our data. Note that our predictions in Fig. 3.1 are based on an isolated vibrational mode. When multiple modes spectrally interfere, the phase contrast is also diminished. Fig. 3.3 illustrates the measured magnitude and phase, and plots real and imaginary components of $\chi^{(2)}$ for the ODT-Au surface, as indicated by the points. Simultaneous fits to the magnitude and phase data resulted in the black lines. When the same set of Lorentzian amplitude, frequency, and width parameters are re-plotted without the non-resonant component, the predicted results are shown in red. From this comparison we can conclude that, although neighboring vibrational modes reduce the phase contrast, the effect is minor in comparison to the effect of the non-resonant contribution. We can see that the methyl symmetric stretch experiences a phase change of nearly $180°$ (red curve in Fig. 3.3b) if $\chi^{(2)}_{\mathrm{NR}} = 0$, in comparison to the measured to phase contrast of ca. $60°$.

3.3.5 Combined use of homodyne and heterodyne SFG data to establish functional group polarity

Although the above examples show that a complete treatment of the heterodyne data will yield the sought polarity information, it is useful to have a manner for extracting this from the homodyne data directly in the case of strong non-resonant contributions that provide clear interference lineshapes. The relationship between all of the experiments and observables is summarized in Fig. 3.4. The left column represents the case of a molecule adsorbed on a surface with no significant non-resonant contribution; the right column presents the same case, but on a metallic substrate with $\phi_{\mathrm{NR}} = -120°$ as in the case of aluminum. The first row depicts the result of a homodyne SFG experiment for a sample with methyl groups directed down (towards the substrate, indicate in blue) and up (away from the substrate, towards the vapor phase, indicated in orange). In the absence of any non-resonant contribution the two spectra are, of course, indistinguishable as seen in Fig. 3.4a. With a large non-resonant contribution (note the extent of the vertical axis in Fig. 3.4b), the two cases of methyl orientations are distinguishable. Practically, this occurs whenever $R > 1$. If a heterodyne experiment were performed on the dielectric substrate,

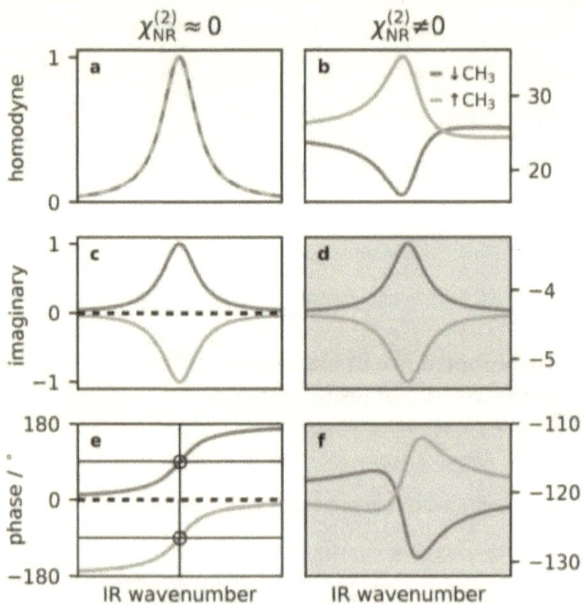

Figure 3.4: Illustration of the relationship between the apparent direction (upwards vs downwards pointing) of a vibrational resonance in homodyne and heterodyne SFG experiments. The first column $\chi_{NR}^{(2)} \approx 0$ represents situations with no non-resonant contribution; the second column shows the same resonance properties with a significant non-resonant background, with $\phi_{NR} = -120°$. Blue spectra are for the methyl symmetric stretch, where the C-to-H axis points towards the substrate. The same vibrational mode, but for the opposite polarity, is shown in orange. The shaded panels represent data that would be useful, but difficult to obtain in a heterodyne experiment. Reprinted with permission from Ref. 123. Copyright 2017 American Chemical Society.

the resonances would appear with opposite signs (Fig. 3.4c) in the extracted imaginary component of the spectra, as we have shown in the case of OTS-glass. Alternatively, and a more direct representation of the measured quantity in the heterodyne experiment, the phase would change by 180°, passing through the resonance at either ±90°, depending on the methyl group orientation (Fig. 3.4e), as encoded in the sign of A. Although the corresponding data in the case of the metal substrate still reveal differences according to the polarity of the methyl group, the imaginary spectra are both negative in this case (Fig. 3.4d), and there is only a small change in phase (Fig. 3.4f) upon passing through the vibrational resonance. The phase contrast depends on the relative magnitudes $R = |\chi_{NR}^{(2)}|/|\chi_{R}^{(2)}|$ and ϕ_{NR}. Figs. 3.4d and 3.4f have been shaded to indicate that, although this information would lead to a determination of the methyl polarity, it is strictly not required, and may be difficult to obtain unless the experimental phase resolution is sufficiently high.

From an inspection of the SDS-Al and ODT-Au data, it is apparent that some phase information is present, but analysis hinges on the non-resonant phase. Furthermore, it has been demonstrated that the non-resonant phase of a metallic substrate may be substantially altered by the presence of the adsorbate, and is sensitive to the surface coverage.[48, 126] However, if heterodyne measurements are available, ϕ_{NR} can be determined experimentally. In the presence of a strong non-resonant contribution, it is then practical and robust to obtain the chemical functional group polarity from the homodyne data. The relationship between the orientation and the direction/appearance of the corresponding resonance feature is given by the sign of $\cos(\phi_{NR} - \phi_R)$ as revealed by Eq. 3.5. If the nature of the SFG signal at that particular IR frequency is not certain, for example if multiple modes may contribute to the observed spectral feature, then both ϕ_{NR} and ϕ_R need to be determined in order to assess this. However, if it is known that the spectral feature is relatively free from contributions of other vibrational modes, a considerable simplification can be made, as was considered in the model in Fig. 3.4, where the phase changes by 180° upon passing through the resonance, and has a value of ±90° on resonance. In this case, we limit $\phi_R = \pm 90°$, and $\cos(\phi_{NR} - \phi_R)$ becomes $\pm \sin \phi_{NR}$.

We define a *polarity parameter P* as

$$P = \text{sgn}(A \cdot \sin \phi_{NR}) \tag{3.7}$$

where $P > 0$ indicates that a functional group pointing up will result in an "upward" resonance in a homodyne measurement. Taking the methyl symmetric stretch in ssp as an example, if the C-to-H axis is directed up, in the direction of the reflected SFG beam, the resonant amplitude in Eq. 3.6 is given by

$$A_{yyz} = N(2\alpha_{ccc}^{(2)} + \alpha_{aac}^{(2)} + \alpha_{bbc}^{(2)})\langle \cos \theta \rangle - N(2\alpha_{ccc}^{(2)} - \alpha_{aac}^{(2)} - \alpha_{bbc}^{(2)})\langle \cos^3 \theta \rangle \tag{3.8}$$

where a, b, c are the molecular frame Cartesian coordinates, with c along the methyl C_3 symmetry axis. In Fig. 3.5, harmonic approximations of these $\alpha^{(2)}$ elements are obtained for methyl groups in different environments. This method has been described in detail previously.[58] In brief, dipole moment vector and polarizability tensor elements (points indicated by open circles) are calculated for various geometries that represent vibration along the normal mode coordinate Q of interest, the methyl symmetric stretch in this case. Fits to second-order polynomials are indicated by solid lines; the slopes evaluated at $Q = 0$ are drawn with dashed lines. In all cases, $\partial \alpha_{aa}/\partial Q \approx \partial \alpha_{bb}/\partial Q$ and so, for the methyl symmetric stretch we can make the further approximation

$$A_{yyz} \approx 2N\frac{\partial \mu_c}{\partial Q}\left[\frac{\partial \alpha_{aa}}{\partial Q} + \frac{\partial \alpha_{cc}}{\partial Q}\right]\langle \cos \theta \rangle + 2N\frac{\partial \mu_c}{\partial Q}\left[\frac{\partial \alpha_{aa}}{\partial Q} - \frac{\partial \alpha_{cc}}{\partial Q}\right]\langle \cos^3 \theta \rangle$$
$$= f_1\langle \cos \theta \rangle + f_3\langle \cos^3 \theta \rangle \tag{3.9}$$

As Fig. 3.5 also reveals that $\partial \alpha_{aa,bb}/\partial Q > 0$, $\partial \alpha_{cc}/\partial Q < 0$, and $\partial \mu_c/\partial Q < 0$, and this results in $f_1 < 0$ and $f_3 < 0$. As a consequence, $A_{yyz} < 0$ when $0° < \theta < 90°$, and $A_{yyz} > 0$ when $90° < \theta < 180°$. On aluminum where $\phi_{NR} = -120°$, $\sin \phi_{NR} \approx -0.87$ and so P is positive. In other words, an "upwards" band (as in the orange spectrum in Fig. 3.4b, and the experimental data in Fig. 3.2g) is predicted. In the case of the terminal methyl group of the ODT akyl chain on gold, $\phi_{NR} = 84°$ and $\sin \phi_{NR} \approx 1$. Examining the ODT-Au

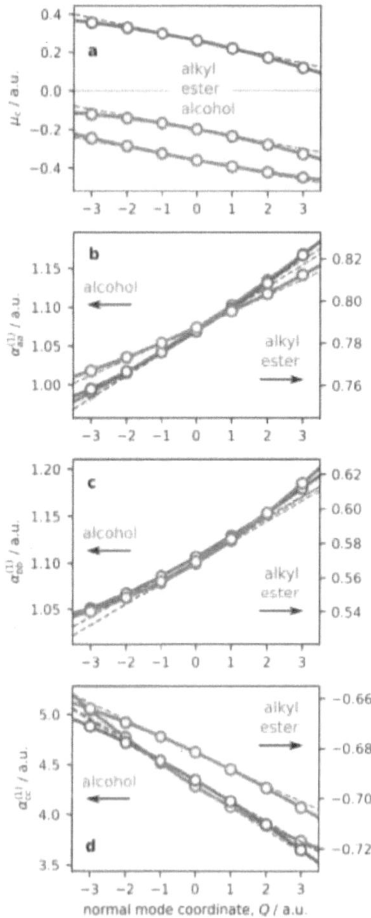

Figure 3.5: The relationship between the sign of the band in the $\mathbb{Im}\{\chi^{(2)}\}$ spectrum, or the direction of the band in the $|\chi^{(2)}|^2$ spectrum in the presence of significant non-resonant background, and the orientation of the chemical functional group depends on the sign of the relevant hyperpolarizability tensor elements. Here we illustrate that, for the methyl symmetric stretch, (a) $\partial\mu_c/\partial Q < 0$, (b,c) $\partial\alpha_{aa}/\partial Q \approx \partial\alpha_{bb}/\partial Q > 0$, and (d) $\partial\alpha_{cc}/\partial Q < 0$, regardless of whether the methyl group is part of an alkyl chain (blue), ester group (green), or alcohol (red). Points are the results of our calculation. Solid lines are the results of a fit to a second-order polynomial; dashed lines are the tangent evaluated at $Q = 0$. Reprinted with permission from Ref. 123. Copyright 2017 American Chemical Society.

homodyne data in Fig. 3.2j, the CH_3 symmetric stretch is downwards pointing. Together with a negative value of P, we again come to the conclusion that the methyls are directed up, away from the gold surface. In the above discussion, we have simplified the consideration of which hyperpolarizability tensor elements in the molecular frame are projected onto the lab frame in the ssp experiment. We also point out that, while the signs of the $\alpha^{(2)}$ elements depend on the choice of molecular axes, the final conclusion does not. If the molecular frame were flipped so the hyperpolarizability elements are now providing a description of the molecule when it is upside down, the resulting tilt angle would now be the opposite quadrant, thereby describing the same molecular orientation. As a second example, if one considers the same hyperpolarizability calculation for the surface water free-OH stretch often observed near 3700 cm^{-1}, one arrives at $A > 0$. Therefore the opposite conclusions are made from experimental data with similar appearance.

Table 3.1 summarizes the workflow for arriving at the polarity of a chemical functional group. In the case of a heterodyne experiment, the direction of $\mathbb{Im}\{\chi^{(2)}\}$ immediately provides the result, as long as the nature of the vibrational mode (from the magnitudes and signs of its relevent hyperpolarizability tensor elements) are known. This is a fundamental requirement for all SFG experiments that seek to establish the functional group polarity, contained in the A term. In the heterodyne experiment, the non-resonant phase only determines the sign of $\mathbb{Im}\{\chi^{(2)}\}$, and this information is irrelevant for our purpose. It is more interesting to consider the homodyne experiment, as the signal-to-noise is high, the measurement is easy and quick to perform, and the analysis is less susceptible to error. Note that the homodyne data we are describing here applies only in the case of a significant non-resonant background with known phase. Here the product of the signs of A and ϕ_{NR} determine the sign of the polarity parameter. A positive value of P indicates that the direction of the peak in the homodyne data indictes the direction of the functional group orientation. If P is negative, the connection is reversed.

Experiment / observation	mode character	NR phase $-180° < \phi < 180°$	polarity parameter	orientation
homodyne "upwards"	$-$	$\phi > 0°$ gold	$-$	$\theta > 90°$ C→H down
	$-$	$\phi < 0°$ aluminum	$+$	$\theta < 90°$ C→H up
	$+$	$\phi > 0°$ gold	$+$	$\theta < 90°$
	$+$	$\phi < 0°$ aluminum	$-$	$\theta > 90°$
homodyne "downwards"	$-$	$\phi > 0°$ gold	$-$	$\theta < 90°$
	$-$	$\phi < 0°$ aluminum	$+$	$\theta > 90°$
	$+$	$\phi > 0°$ gold	$+$	$\theta > 90°$
	$+$	$\phi < 0°$ aluminum	$-$	$\theta < 90°$
heterodyne "upwards"	$-$	not required		$\theta > 90°$
	$+$			$\theta < 90°$
heterodyne "downwards"	$-$			$\theta < 90°$
	$+$			$\theta > 90°$

Table 3.1: Scheme for determining the polarity of a chemical function group based on either a homodyne measurement (when there is significant non-resonant amplitude), or a heterodyne measurement. In all cases, a coincident external reflection geometry is assumed, where methyls pointing up have their C-to-H vector parallel to the reflected SFG beam. Reprinted with permission from Ref. 123. Copyright 2017 American Chemical Society.

3.4 Conclusions

Phase-resolved SFG spectroscopy is a powerful tool for elucidating the structure of molecules on surfaces that is uniquely capable of resolving the polarity of each chemical group orientation, i.e. distinguishing the quadrants that define the up vs down directions. This technique has been especially valued for the study of dielectric surfaces, where the lack of a non-resonant contribution offers few other clues to the polarity, as interference effects are generally weak or non-existent. On metal surfaces, phase-resolved experiments also provide information of the surface electronic structure, which is in turn sensitive to the nature and coverage of adsorbed species. However, there are practical challenges associated with measuring the resonant phase on metal surfaces, as the phase contrast is observed (and anticipated) to be low. One solution to this problem is to use the phase of the metal (preferably in the presence of the absorbed molecules) as determined in a heterodyne scheme, together with the homodyne data, in order to extract the sought polarity, as an alternative to retrieving the complete imaginary spectrum.

Chapter 4

Determining Nonlinear Optical Coefficients of Metals by Multiple Angle of Incidence Heterodyne-Detected Sum-Frequency Generation Spectroscopy[*,†]

4.1 Introduction

Visible-infrared sum-frequency generation is a second-order nonlinear vibrational spectroscopy technique that is valued for its ability to probe the structure of molecules at solid, liquid, and vapor interfaces.[127–129] Under the electric dipole approximation, the second-order electric susceptibility $\chi^{(2)}$ is non-zero only in the absence of inversion symmetry.[40,63] Therefore, when a broadband or tunable infrared beam is spatially- and temporally-overlapped with a fixed frequency visible or near-infrared beam, vibrational resonances in $\chi^{(2)}$ of surface species may be observed. Performing the experiment with different beam polarizations enables different elements of the rank-3 $\chi^{(2)}$ tensor to be obtained. If the orientation of a chemical functional group is of interest, measurements

---

[*]Reproduced in part from Yang, W.-C.; Busson, B.; Hore, D.K. "Determining Nonlinear Optical Coefficients of Metals by Multiple Angle of Incidence Heterodyne-Detected Sum-Frequency Spectroscopy." *J. Chem. Phys.* **152**, 084708 (2020). Copyright 2020 American Institute of Physics.

[†]All sample preparation, data collection, data analysis, and multi-angle model development by Wei-Chen Yang. Theoretical development related to the quadrupolar components of $\chi^{(2)}$ contributed by Bertrand Busson.

are typically carried out in at least two polarization schemes. For example, an experiment in which the s-polarized component of the SFG is measured when s-polarized visible and p-polarized infrared is used (the so-called ssp scheme) selectively probes $\chi^{(2)}_{yyz}$ (see Fig. 4.1). Similarly, sps experiments provide access to $\chi^{(2)}_{yzy}$. If we consider the methyl symmetric stretch, the orientation of its C_3 axis may be determined from the ratio $\chi^{(2)}_{yyz}/\chi^{(2)}_{yzy}$. There are many descriptions in the literature that provide details on this procedure.[27, 54, 130–134]

Collecting spectra in multiple polarization schemes generally works well for dielectric interfaces, keeping in mind that the surface fields are enhanced or reduced at specific angles of incidence, and the optimum settings depend on the polarization of all beams.[75, 135, 136] In the case of metals, however, the amplitude of s-polarized light (electric field vector perpendicular to the plane of incidence) at the surface is strongly reduced. The extent to which the surface field decreases is dependent on the metal and specific frequency, but often creates the situation where only the ppp polarization scheme yields appreciable signals. This presents a challenge for orientation analysis as $\chi^{(2)}_{xxz}$, $\chi^{(2)}_{xzx}$, $\chi^{(2)}_{zxx}$ and $\chi^{(2)}_{zzz}$ are all potential contributors to the measured response. If the nature of the surface species and the vibrational assignment are well understood, then one option is to perform homodyne (intensity) measurements of ppp spectra at multiple angles of incidence and then simultaneously solve for the amplitudes of interest by assuming a specific lineshape, such as a Lorentzian.[38, 39, 56, 80] For samples that generate response with both s- and p-components of the incident fields, there are methods that can extract the $\chi^{(2)}$ tensor elements by varying the beam polarizations.[137, 138]

Here we present an approach that is universally applicable, can measure complex-valued $\chi^{(2)}$ elements, and is able to provide the dispersion of these quantities throughout the region over which the laser frequencies are tuned. We demonstrate our method applied to the case of the gold surface, of interest to many surface studies.[44, 115, 116, 124, 125] The choice of gold is further motivated by its universal use as a support for (e.g. thiolated) self-assembled molecular layers, its chemical stability in air and its intense SFG nonresonant response (due to contributions from its free and bound electrons.[139, 140]) We begin by introducing some formalism that relates the surface fields to the incidence and reflected SFG fields. We also describe the quadrupolar response from the bulk of the material. We then analyze heterodyne-detected ssp SFG data to obtain the magnitude and phase of $\chi_{yyz}^{(2)}$ from a gold surface when the visible laser is fixed at 532 nm, and the infrared beam has a frequency of 2800 cm^{-1}. We use these results to verify the outcome of our analysis of multi-angle heterodyne ppp experiment on gold from 40–70°. In the end, our method is applicable to any dielectric or metal surface, but is particularly valuable when structural information is sought and two or three independent $\chi^{(2)}$ elements must be extracted from ppp data.

4.2 Background

4.2.1 Nonlinear optical response of metal surface and bulk in SFG

Before turning to the description of our heterodyne angle-dependent method, we first review the essential nonlinear properties of gold in order to determine which tensor components have to be taken into account in the data analysis. The second-order nonlinear optical response of a metal surface has been studied soon after the first reported evidence of second harmonic generation (SHG),[141–143] and continues to be of interest in the SHG and SFG community.[144–152] The description of the nonlinear gold response presented here follows from the general phenomenological model of surface and bulk SFG response,[139, 141, 153–155] analogous to SHG models, such as that developed by Mizrahi and Sipe.[156] For a polycrystalline cubic material, the bulk and surface behave essentially like an isotropic system. The surface second-order polarization has its origins in the discontinuity of the electron density while crossing the interface at $z = 0$ (see Fig. 4.1), creating large electric field gradients.[154, 155] The resulting surface nonlinear polarization is obtained by integration of the bulk polarization along z across the interface.[153] The surface terms of the gold second-order susceptibility $\chi_{ijk,\text{surf}}^{(2)}$, defined as $P_i^S = \chi_{ijk,\text{surf}}^{(2)} E_{\text{vis,j}} E_{\text{IR,k}}$ have $C_{\infty v}$

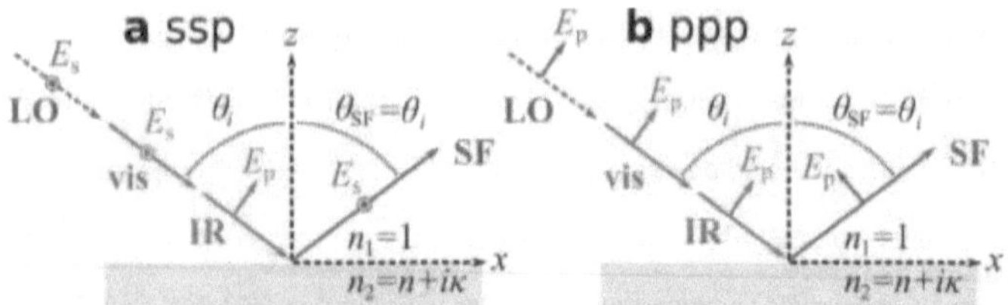

Figure 4.1: Definition of the coordinate system and conventions used in this analysis for the (a) ssp and (b) ppp polarization schemes. z is along the surface normal, (x, y) define the interface between air and gold, and (x, z) is the plane of incidence. Reprinted with permission from Ref. 158. Copyright 2020 American Institute of Physics.

symmetry, leaving only the usual seven non-vanishing terms zzz, $xxz = yyz$, $xzx = yzy$ and $zxx = zyy$ (Table 4.1).[29] The parallel components of the electric fields experience no gradients as they are continuous across the interface, so the surface zxx term vanishes.[155, 157]

Apart from these surface-specific contributions, the emitted SFG also contains contributions from bulk effects, which we present in more detail as their properties fundamentally differ from the well-known surface terms. In isotropic materials, they arise from the gradients associated with the propagation of the light waves (electric fields $\mathbf{E}_{vis}$ and $\mathbf{E}_{IR}$) inside the bulk, and therefore depend on their bulk wavevectors $\mathbf{q}_{vis}$ and $\mathbf{q}_{IR}$ where $\mathbf{q}_i = q_{i,x}\hat{x} - q_{i,z}\hat{z}$ and

$$q_{i,x} = \frac{\omega_i}{c}\sin\theta_i = \frac{n_i\omega_i}{c}\sin\theta_i^T,$$
$$q_{i,z} = n_i\frac{\omega_i}{c}\cos\theta_i^T, \tag{4.1}$$

where n_i is the bulk refractive index at ω_i (refractive index of air is taken as unity), and θ_i and θ_i^T are the angles of incidence of beam i in air and in the bulk, respectively. The situation is different when bulk dipolar contributions are allowed, for example in a chiral liquid,[159, 160] as they will dominate the quadrupolar response due to the gradients. The penetration of light inside the gold is limited by the large value of the attenuation coefficient: at 532 nm, the electric field decays to $1/e$ of its initial value after 37 nm; at 2800 cm^{-1} the corresponding value is 23 nm. However, considering fcc gold with a lattice constant around 0.4 nm, the penetration depth is greater than 50 unit cells. The bulk terms are therefore important to consider. In addition, large attenuation results in large wavevectors in Eq. 4.1, leading to large field gradients as will be discussed later (Eq. 4.3).

In the literature, the comparison between surface and bulk terms has been extensively studied for SHG,[161,162] but less so for SFG.[25,40,163–168] There are two ways to describe the bulk contributions from an isotropic medium. The first one is inspired from the original work of Bloembergen and Pershan,[141] separating the nonlinear polarization inside the bulk into components parallel and perpendicular to the source wavevector $\mathbf{q}_{vis} + \mathbf{q}_{IR}$. Each component radiates in the reflected and transmitted directions as deduced from the boundary conditions at the interface. This formulation leads to a single, compact but intricate, bulk contribution to the effective nonlinear susceptibility that is added to the surface terms.[139,154,155] In the isotropic case, the bulk source is parametrized by the non-vanishing coefficients of the quadrupolar nonlinear susceptibility defined by

$$\mathbf{P}^B(\mathbf{q}_{vis},\mathbf{q}_{IR},\mathbf{r}) = \mathbf{P}^B(\mathbf{q}_{vis},\mathbf{q}_{IR})\exp\left[i(\mathbf{q}_{vis}+\mathbf{q}_{IR})\cdot\mathbf{r}\right] \tag{4.2}$$

with

$$\begin{aligned}\mathbf{P}^B(\mathbf{q}_{vis},\mathbf{q}_{IR}) = i\,[&D_{vis}\mathbf{q}_{vis}(\mathbf{E}_{vis}\cdot\mathbf{E}_{IR}) + \Delta_{vis}(\mathbf{q}_{vis}\cdot\mathbf{E}_{IR})\mathbf{E}_{vis}\\ &+D_{IR}\mathbf{q}_{IR}(\mathbf{E}_{vis}\cdot\mathbf{E}_{IR}) + \Delta_{IR}(\mathbf{q}_{IR}\cdot\mathbf{E}_{vis})\mathbf{E}_{IR}]\end{aligned} \tag{4.3}$$

under the plane wave hypothesis $\mathbf{q}_i \cdot \mathbf{E}_i = 0$. Analogous coefficients have long been defined in SHG[169,170] and extension to crystalline cubic materials is possible by adding anisotropic terms.[171,172] A quantitative evaluation of these coefficients is necessary to determine the impact of the bulk response as compared to the surface terms. For this purpose, simple models enable calculation of the D_i and Δ_i coefficients in Eq. 4.3 from the electronic properties of gold,[155] while a more rigorous description requires separating the contributions from free and bound electrons.[139,173] Surface and bulk effective susceptibilities, calculated separately, reconstruct the total SFG response from gold.[140]

The second formulation for the bulk terms describes the bulk as an infinite stack of thin plates, just like the surface terms. Considering that the bulk polarization in Eq. 4.2 is independent of the position except for its phase term, each plate radiates the same electric field as if it were located at $z = 0$, taking into account the depth-dependent phase shift $\Delta\phi = (\Delta\mathbf{q})_R \cdot \mathbf{r}$ where $(\Delta\mathbf{q})_R = \mathbf{q}_{vis} + \mathbf{q}_{IR} - \mathbf{q}_{SFG}$, with $\mathbf{q}_{SFG} = q_{SFG,x}\hat{x} + q_{SFG,z}\hat{z}$ in reflection. Because of momentum conservation parallel to the surface,[141] the wavevector mismatch $\Delta\mathbf{q}_R$ applies to only the z-components of the wavevectors and $\Delta\mathbf{q}_R = -(\Delta q_z)_R\,\hat{z}$. Integration along z produces an equivalent surface contribution defined by[40,165]

$$\mathbf{P}^{B,\mathrm{surf}}(\mathbf{q}_{vis}, \mathbf{q}_{IR}) = \frac{i\,\mathbf{P}^B(\mathbf{q}_{vis}, \mathbf{q}_{IR})}{(\Delta q_z)_R} \tag{4.4}$$

to which the formulas usually devoted to surface terms may be applied. In particular, an equivalent surface nonlinear susceptibility of the bulk is defined as

$$P_i^{B,\mathrm{surf}}(\mathbf{q}_{vis}, \mathbf{q}_{IR}) = \chi^{(2)}_{ijk,\mathrm{bulk}}(\mathbf{q}_{vis}, \mathbf{q}_{IR})E_{j,vis}E_{k,IR}. \tag{4.5}$$

Comparing SFG emitted in reflection and transmission, in principle, enables the bulk and surface contributions to be separated.[167] It has been demonstrated that part of this bulk nonlinear susceptibility may be transformed into a true surface nonlinear susceptibility $\chi^{(2)}_{ijk,\mathrm{bulk,insep.}}$, making it experimentally indistinguishable from $\chi^{(2)}_{ijk,\mathrm{surf}}$.[40,165]

Both formulations have their unique advantages, the first one facilitating modeling of the dispersion of the nonlinear bulk coefficients, and the second one providing a more facile route for comparing and combining bulk and surface terms. However, at first sight they may seem incompatible because the formulas for the emitted field in the bulk have diverged (compare for example Refs. 155 and 40). We show in Appendix A that they are in fact equivalent; we may therefore directly express $\chi^{(2)}_{ijk,\mathrm{bulk}}$ as a function of the D_i and Δ_i coefficients to calculate the bulk contribution. Care must be taken when describing the separable and inseparable bulk nonlinear susceptibilities, as the separable part of this equivalent surface nonlinear susceptibility only formally behaves as a true surface nonlinear susceptibility as far as SFG emission is concerned, but essentially differs through

tensor element	polarization	dipolar surface term	inseparable term	separable bulk (non-separable as a function of angle)	separable bulk (separable as a function of angle)
xxx	ppp	-	-	-	yes (1.1)
xzz	ppp	-	-	-	yes (0.008)
zzx	ppp	-	-	-	yes (0.008)
zxz	ppp	-	-	-	yes (0.004)
xxz	ppp	yes	-	yes (0.007)	-
xzx	ppp	yes	-	yes (3.0)	-
zxx	ppp	yes (but 0)	yes	-	yes (0.48)
zzz	ppp	yes	yes	yes (0.03)	-
yyx	ssp	-	-	yes (0.005)	-
yyz	ssp	yes	-	yes (0.005)	-
zyy	pss	yes (but 0)	yes	-	yes (0.31)
xyy	pss	-	-	-	yes (0.44)
yxy	sps	-	-	-	yes (0.2)
yzy	sps	yes	-	yes (2.0)	-

Table 4.1: All 14 possible non-zero elements of $\chi^{(2)}$, grouped according to the polarization scheme in which they are probed.

its wavevector ($\mathbf{q}_{vis}$, $\mathbf{q}_{IR}$ and $\Delta\mathbf{q}$) dependence. As a consequence, the usual symmetry considerations do not apply to this tensor and a total of 14 terms is expected, listed in Table 4.1. As detailed in Appendix B, a subset of these coefficients are intrinsically inseparable from the zxx, zyy and zzz surface terms. In the following, we focus on the separable contributions (Eqs. 4.28–4.31 and 4.35–4.38).

Our goal is to tune the angle of incidence θ to separate the various contributions to the gold SFG response as a result of their specific angular dependence. Among the bulk separable terms of the equivalent nonlinear surface susceptibility, those that do not vary with θ behave exactly like a surface term in our experiment. Following our previous work,[139] we have calculated them for both the free and bound electron populations of gold, and display their magnitudes in Fig. 4.2. We note that the most intense terms (xzx, yzy and zzz) are the same for both electron populations (this is also true for all other terms except xxz, yyz, zzx and yyx) and vary by no more than a few percent over the considered angle range, as do xxz and yyz. These five elements cannot be separated from surface contributions in our measurements (see Table 4.1). Therefore, in order to estimate the most important terms among the nine remaining ones, we calculate the total effective

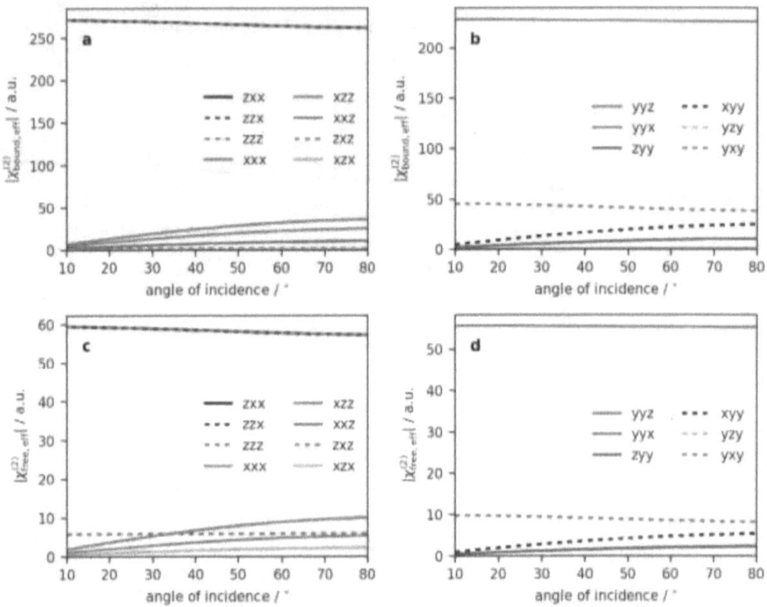

Figure 4.2: Separable $|\chi^{(2)}_{\text{bulk}}|$ as a function of angle of incidence, dividing the contribution into that from the (a, b) the interband transition and (c, d) the free electron contribution. Reprinted with permission from Ref. 158. Copyright 2020 American Institute of Physics.

contribution by summing free and bound electron terms (although their mutual weights may be debated[140]) and weighting them by the appropriate Fresnel coefficients (Fig. 4.3). Their relative magnitude is indicated in the last column of Table 4.1.

Maximal compatibility between bulk and surface terms is ensured when all quantities involved (electric fields and polarizations) are defined in the same medium, chosen as the place where the nonlinearity effectively takes place. In the present case, we consider this location to be inside gold, at $z = 0^-$.

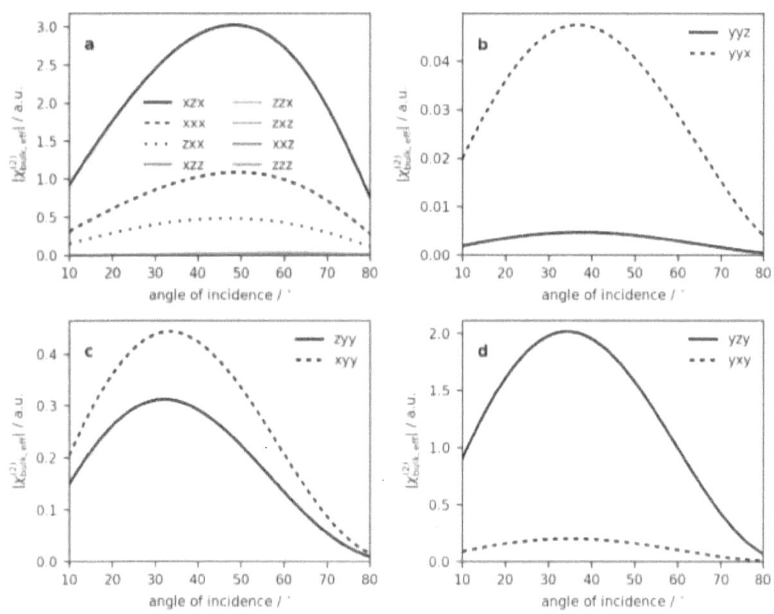

Figure 4.3: Bulk effective $\chi^{(2)}$ elements as a function of angle of incidence for all four polarization schemes (a) PPP, (b) SSP, (c) PSS, and (d) SPS. The xzz, zzx, zxz, xxz, and zzz elements that contribute to PPP are too small to be seen in comparison to xzx, xxx, zxx. Reprinted with permission from Ref. 158. Copyright 2020 American Institute of Physics.

4.2.2 Local field considerations

The intensity of the reflected SFG is obtained from

$$I_{\text{SFG}} = \frac{8\pi^3 \omega_{\text{SFG}}^2}{c^3 \cos^2 \theta_{\text{SFG}}} |\chi_{\text{eff}}^{(2)}|^2 I_{\text{vis}} I_{\text{IR}} \tag{4.6}$$

where the effective susceptibility is defined as

$$\chi_{\text{eff}}^{(2)} = L_{ii,\text{SFG}} L_{jj,\text{vis}} L_{kk,\text{IR}} e_{i,\text{SFG}} e_{j,\text{vis}} e_{k,\text{IR}} \left(\chi_{ijk,\text{surf}}^{(2)} + \chi_{ijk,\text{bulk}}^{(2)} \right). \tag{4.7}$$

This definition includes elements of the unit polarization vector

$$\mathbf{e} = \begin{bmatrix} \pm\cos\theta \\ 1 \\ \sin\theta \end{bmatrix} \tag{4.8}$$

and local field correction tensor

$$\mathbf{L} = \begin{bmatrix} 1 - r_p & 0 & 0 \\ 0 & 1 + r_s & 0 \\ 0 & 0 & (1 + r_p)\dfrac{1}{n^2} \end{bmatrix} \tag{4.9}$$

with r_p and r_s the familiar Fresnel reflection coefficients

$$r_p = \frac{n\cos\theta - \cos\theta^{\text{T}}}{\cos\theta^{\text{T}} + n\cos\theta} \tag{4.10a}$$

$$r_s = \frac{\cos\theta - n\cos\theta^{\text{T}}}{\cos\theta + n\cos\theta^{\text{T}}} \tag{4.10b}$$

where the refracted angle θ^{T} is obtained from Snell's law. As a reminder, we have assumed the refractive index is unity for air, and n refers to the complex refractive index of bulk gold in all of our expressions. In our experimental geometry (Fig. 4.1), positive values of e_x are used to describe in the incoming visible and IR beams; a negative value of e_x is used in the case of the reflected SFG beam. For the vibrationally non-resonant response of gold, the form of L_{zz} indicates that we treat the signal as if it originates from the bulk gold just below the surface (at $z = 0^-$).

The relative magnitude and angle dependence of the $\mathbf{L}$ factors applicable to the ppp polarization scheme is shown in Fig. 4.4 for the case air–gold interface. In the case where

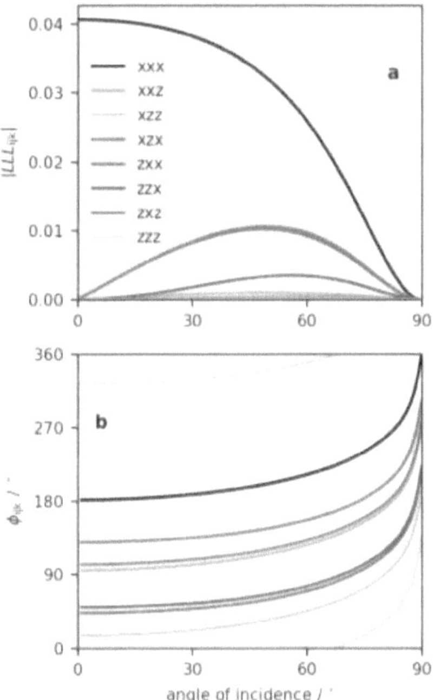

Figure 4.4: A comparison angle-dependent (a) magnitude and (b) phase of the local field correction factors for the air–gold interface. Reprinted with permission from Ref. 158. Copyright 2020 American Institute of Physics.

the refractive index of the visible and SFG beams is similar, and the angle of the reflected SFG is close to the angle of incident visible beam (they are equal in our collinear geometry) we expect $L_{xx,\text{SFG}}L_{zz\text{vis}} = L_{zz,\text{SFG}}L_{xx,\text{vis}}$.

4.2.3 Bulk SFG contribution transformed into a surface term

We recall[141] that, inside the bulk, the SFG beam propagates with a wavevector $\mathbf{q}_{\text{SFG}}$, a dielectric function ε_{SFG} and at an angle of incidence $\theta_{\text{SFG}}^{\text{T}}$, whereas the nonlinear polarization propagates with a wavevector $\mathbf{q}_{\text{SFG}}^{S} = \mathbf{q}_{\text{vis}} + \mathbf{q}_{\text{IR}}$, associated to a dielectric function $\varepsilon_S = n_S^2 = \varepsilon_{\text{SFG}}|\mathbf{q}_{\text{SFG}}^{S}|^2/|\mathbf{q}_{\text{SFG}}|^2$ at an angle of incidence θ_{SFG}^{S}. Starting from the original equations giving the amplitudes of the reflected SFG field as a function of the bulk polarization[154]

$$E_{\mathrm{p}}(\omega_{\mathrm{SFG}}) = \frac{4\pi}{n_S n_{\mathrm{SFG}}} \frac{1}{n_{\mathrm{SFG}}\cos\theta_{\mathrm{SFG}} + \cos\theta_{\mathrm{SFG}}^T} \left[P_{\parallel}^{\mathrm{B}} \sin\theta_{\mathrm{SFG}} \right.$$
$$\left. + P_{\perp}^{\mathrm{B}} \frac{\varepsilon_{\mathrm{SFG}} n_S \cos\theta_{\mathrm{SFG}}^S - \varepsilon_S n_{\mathrm{SFG}} \cos\theta_{\mathrm{SFG}}^T}{\varepsilon_{\mathrm{SFG}} - \varepsilon_S} \right] \tag{4.11}$$

and

$$E_{\mathrm{s}}(\omega_{\mathrm{SFG}}) = E_{\mathrm{y}}(\omega_{\mathrm{SFG}}) = \frac{4\pi}{\varepsilon_{\mathrm{SFG}} - \varepsilon_S} \frac{n_S \cos\theta_{\mathrm{SFG}}^S - n_{\mathrm{SFG}}\cos\theta_{\mathrm{SFG}}^T}{n_{\mathrm{SFG}}\cos\theta_{\mathrm{SFG}}^T + \cos\theta_{\mathrm{SFG}}} P_y^{\mathrm{B}} \tag{4.12}$$

where $P_{\parallel}^{\mathrm{B}}$ and $P_{\perp}^{\mathrm{B}}$ stand for the components of the bulk nonlinear polarization parallel and perpendicular to $\mathbf{q}_{\mathrm{SFG}}^S$, respectively, in the (x,z) plane. Defining the wavevector mismatch in reflection $(\Delta q_z)_R = q_{\mathrm{vis},z} + q_{\mathrm{IR},z} + q_{\mathrm{SFG},z}$ and in transmission $(\Delta q_z)_T = q_{\mathrm{vis},z} + q_{\mathrm{IR},z} - q_{\mathrm{SFG},z}$, we use the following identities to transform Eq. 4.11 and 4.12

$$(\mathbf{q}_{\mathrm{SFG}}^S)^2 - (\mathbf{q}_{\mathrm{SFG}})^2 = (q_{\mathrm{SFG},z}^S)^2 - (q_{\mathrm{SFG},z})^2$$
$$= (\Delta q_z)_R (\Delta q_z)_T \tag{4.13}$$
$$= \frac{\omega_{\mathrm{SFG}}^2}{c^2}(\varepsilon_S - \varepsilon_3)$$

$$(\Delta q_z)_{R/T} = \frac{\omega_{\mathrm{SFG}}}{c}(n_S \cos\theta_{\mathrm{SFG}}^S \pm n_{\mathrm{SFG}}\cos\theta_{\mathrm{SFG}}^T). \tag{4.14}$$

For s-polarization, this leads to

$$E_{\mathrm{s}}(\omega_{\mathrm{SFG}}) = \frac{4i\pi\omega_{\mathrm{SFG}}}{c} \frac{1}{n_{\mathrm{SFG}}\cos\theta_{\mathrm{SFG}}^T + \cos\theta_{\mathrm{SFG}}} \frac{iP_y^{\mathrm{B}}}{(\Delta q_z)_R}$$
$$= \frac{2i\pi\omega_{\mathrm{SFG}}}{\cos\theta_{\mathrm{SFG}} c} L_{yy}(\omega_{\mathrm{SFG}}) \frac{iP_y^{\mathrm{B}}}{(\Delta q_z)_R} \tag{4.15}$$

and for p-polarization we have

$$P_{\parallel}^{\mathrm{B}} = P_x^{\mathrm{B}} \sin\theta_{\mathrm{SFG}}^S - P_z^{\mathrm{B}} \cos\theta_{\mathrm{SFG}}^S$$
$$P_{\perp}^{\mathrm{B}} = P_x^{\mathrm{B}} \cos\theta_{\mathrm{SFG}}^S + P_z^{\mathrm{B}} \sin\theta_{\mathrm{SFG}}^S. \tag{4.16}$$

Using

$$\frac{\varepsilon_{\mathrm{SFG}} n_S \cos\theta_{\mathrm{SFG}}^S - \varepsilon_S n_{\mathrm{SFG}}\cos\theta_{\mathrm{SFG}}^T}{\varepsilon_{\mathrm{SFG}} - \varepsilon_S} = n_{\mathrm{SFG}}\left(\cos\theta_{\mathrm{SFG}}^T - \frac{\omega_{\mathrm{SFG}}}{c} \frac{n_{\mathrm{SFG}}}{(\Delta q_z)_R}\right) \tag{4.17}$$

we get, for the x projection,

$$E_{\text{p},x}(\omega_{\text{SFG}}) = \frac{4\pi}{n_S n_{\text{SFG}}} \frac{1}{n_{\text{SFG}}\cos\theta_{\text{SFG}} + \cos\theta_{\text{SFG}}^T} \Big[\sin\theta_{\text{SFG}}^S \sin\theta_{\text{SFG}}$$
$$+ \cos\theta_{\text{SFG}}^S n_{\text{SFG}}(\cos\theta_{\text{SFG}}^T - \frac{\omega_{\text{SFG}}}{c}\frac{n_{\text{SFG}}}{(\Delta q_z)_R})\Big] P_x^{\text{B}}. \tag{4.18}$$

Using $\sin\theta_{\text{SFG}} = n_{\text{SFG}}\sin\theta_{\text{SFG}}^T$ and

$$(\Delta q_z)_R \cos(\theta_{\text{SFG}}^T - \theta_{\text{SFG}}^S) = \frac{\omega_{\text{SFG}}}{c}\Big[n_{\text{SFG}}\cos\theta_{\text{SFG}}^S + n_S\cos\theta_{\text{SFG}}^T\Big], \tag{4.19}$$

this simplifies into

$$E_{\text{p},x}(\omega_{\text{SFG}}) = \frac{4\pi\omega_{\text{SFG}}}{c} \frac{\cos\theta_{\text{SFG}}^T}{n_{\text{SFG}}\cos\theta_{\text{SFG}} + \cos\theta_{\text{SFG}}^T} \Big[\frac{P_x^{\text{B}}}{(\Delta q_z)_R}\Big]$$
$$= -\frac{2i\pi\omega_{\text{SFG}}}{\cos\theta_{\text{SFG}}c} L_{xx}(\omega_{\text{SFG}})\cos\theta_{\text{SFG}} \frac{iP_x^{\text{B}}}{(\Delta q_z)_R}. \tag{4.20}$$

For the z projection, we have

$$E_{\text{p},z}(\omega_{\text{SFG}}) = \frac{4\pi}{n_S n_{\text{SFG}}} \frac{1}{n_{\text{SFG}}\cos\theta_{\text{SFG}} + \cos\theta_{\text{SFG}}^T} \Big[-\cos\theta_{\text{SFG}}^S \sin\theta_{\text{SFG}}$$
$$+ \sin\theta_{\text{SFG}}^S n_{\text{SFG}}(\cos\theta_{\text{SFG}}^T - \frac{\omega_{\text{SFG}}}{c}\frac{n_{\text{SFG}}}{(\Delta q_z)_R})\Big] P_z^{\text{B}}, \tag{4.21}$$

simplifying by use of

$$(\Delta q_z)_R \sin(\theta_{\text{SFG}}^T - \theta_{\text{SFG}}^S) = \frac{\omega_{\text{SFG}}}{c}\frac{\varepsilon_S - \varepsilon_{\text{SFG}}}{n_S n_{\text{SFG}}} \sin\theta_{\text{SFG}} \tag{4.22}$$

into

$$E_{\text{p},z}(\omega_{\text{SFG}}) = -\frac{4\pi\omega_{\text{SFG}}}{c\,n_{\text{SFG}}} \frac{1}{n_{\text{SFG}}\cos\theta_{\text{SFG}} + \cos\theta_{\text{SFG}}^T} \frac{P_z^{\text{B}}}{(\Delta q_z)_R} \sin\theta_{\text{SFG}}$$
$$= \frac{2i\pi\omega_{\text{SFG}}}{c\cos\theta_{\text{SFG}}} L_{zz}(\omega_{\text{SFG}})\sin\theta_{\text{SFG}} \frac{iP_z^{\text{B}}}{(\Delta q_z)_R}. \tag{4.23}$$

Equations 4.15, 4.20 and 4.23 are formally identical to the decomposition of the electric field emitted at the interface by an equivalent surface polarization[153] defined by Eq. 4.4.

4.2.4 Separable and inseparable bulk terms

From Eq. 4.3, 4.4 and 4.5, and using Eq. 4.1, we get for the 14 components of the equivalent surface nonlinear susceptibility:

$$\chi_{xxx,\text{bulk}}^{(2)} = -\frac{1}{(\Delta q_z)_R}\Big[D_{\text{vis}}q_{\text{vis},x} + D_{\text{IR}}q_{\text{IR},x} + \Delta_{\text{vis}}q_{\text{vis},x} + \Delta_{\text{IR}}q_{\text{IR},x}\Big] \tag{4.24}$$

$$\chi^{(2)}_{xzz,\text{bulk}} = \chi^{(2)}_{xyy,\text{bulk}} = -\frac{1}{(\Delta q_z)_R}\left[D_{\text{vis}}q_{\text{vis},x} + D_{\text{IR}}q_{\text{IR},x}\right] \tag{4.25}$$

$$\chi^{(2)}_{zxx,\text{bulk}} = \chi^{(2)}_{zyy,\text{bulk}} = \frac{1}{(\Delta q_z)_R}\left[D_{\text{vis}}q_{\text{vis},z} + D_{\text{IR}}q_{\text{IR},z}\right] \tag{4.26}$$

$$\chi^{(2)}_{zzz,\text{bulk}} = \frac{1}{(\Delta q_z)_R}\left[D_{\text{vis}}q_{\text{vis},z} + D_{\text{IR}}q_{\text{IR},z} + \Delta_{\text{vis}}q_{\text{vis},z} + \Delta_{\text{IR}}q_{\text{IR},z}\right] \tag{4.27}$$

$$\chi^{(2)}_{xxz,\text{bulk}} = \chi^{(2)}_{yyz,\text{bulk}} = \frac{1}{(\Delta q_z)_R}\left[\Delta_{\text{vis}}q_{\text{vis},z}\right] \tag{4.28}$$

$$\chi^{(2)}_{xzx,\text{bulk}} = \chi^{(2)}_{yzy,\text{bulk}} = \frac{1}{(\Delta q_z)_R}\left[\Delta_{\text{IR}}q_{\text{IR},z}\right] \tag{4.29}$$

$$\chi^{(2)}_{zzx,\text{bulk}} = \chi^{(2)}_{yyx,\text{bulk}} = -\frac{1}{(\Delta q_z)_R}\left[\Delta_{\text{vis}}q_{\text{vis},x}\right] \tag{4.30}$$

$$\chi^{(2)}_{zxz,\text{bulk}} = \chi^{(2)}_{yxy,\text{bulk}} = -\frac{1}{(\Delta q_z)_R}\left[\Delta_{\text{IR}}q_{\text{IR},x}\right], \tag{4.31}$$

where the bulk response coefficients, D and Δ are described in detail in the Appendix of Ref. 139. It is known[153] that only the transverse part of the bulk nonlinear polarization emits field at the SFG frequency, the transversality being related to its wavevector $\mathbf{q}_{\text{SFG}}$. We rewrite Eq. 4.3 as

$$\begin{aligned}
-i\mathbf{P}^{\mathbf{B}}(\mathbf{q}_{\text{vis}}, \mathbf{q}_{\text{IR}}) = &\left[\frac{D_{\text{vis}} + D_{\text{IR}}}{2}(\mathbf{q}_{\text{vis}} + \mathbf{q}_{\text{IR}})\right. \\
&\left. + \frac{D_{\text{vis}} - D_{\text{IR}}}{2}(\mathbf{q}_{\text{vis}} - \mathbf{q}_{\text{IR}})\right](\mathbf{E}_{\text{vis}} \cdot \mathbf{E}_{\text{IR}}) \\
&+ \Delta_{\text{vis}}(\mathbf{q}_{\text{vis}} \cdot \mathbf{E}_{\text{IR}})\mathbf{E}_{\text{vis}} + \Delta_{\text{IR}}(\mathbf{q}_{\text{IR}} \cdot \mathbf{E}_{\text{vis}})\mathbf{E}_{\text{IR}}
\end{aligned} \tag{4.32}$$

We may, without modifying the emitted fields, subtract from the first term a non-radiative contribution $(D_{\text{vis}} + D_{\text{IR}})\mathbf{q}_{\text{SFG}}(\mathbf{E}_{\text{vis}} \cdot \mathbf{E}_{\text{IR}})/2$, so that it now contributes to the bulk nonlinear polarization as

$$-i\frac{D_{\text{vis}} + D_{\text{IR}}}{2}(\Delta q_z)_R(\mathbf{E}_{\text{vis}} \cdot \mathbf{E}_{\text{IR}})\hat{z} \tag{4.33}$$

and to the equivalent surface polarization as

$$\mathbf{P}^{\text{B,surf,insep.}} = \frac{D_{\text{vis}} + D_{\text{IR}}}{2}(\mathbf{E}_{\text{vis}} \cdot \mathbf{E}_{\text{IR}})\hat{z}. \tag{4.34}$$

This part of the equivalent surface polarization amounts to adding the constant value $(D_{\text{vis}} + D_{\text{IR}})/2$ to the surface coefficients zxx, zyy and zzz. As it does not depend on any wavevector anymore, it is not possible to experimentally discriminate it from pure surface terms,[171] and is named inseparable bulk.[167] On the contrary, the separable bulk terms of the equivalent surface nonlinear polarization follow, modifying Eqs. 4.24–4.27 to obtain

$$\chi^{(2)}_{xxx,\text{bulk,sep.}} = -\frac{1}{(\Delta q_z)_R}\left[\frac{D_{\text{vis}}-D_{\text{IR}}}{2}(q_{\text{vis},x}-q_{\text{IR},x})+\Delta_{\text{vis}}q_{\text{vis},x}+\Delta_{\text{IR}}q_{\text{IR},x}\right] \tag{4.35}$$

$$\begin{aligned}\chi^{(2)}_{xzz,\text{bulk,sep.}} &=\chi^{(2)}_{xyy,\text{bulk,sep.}}\\ &=-\frac{1}{(\Delta q_z)_R}\left[\frac{D_{\text{vis}}-D_{\text{IR}}}{2}(q_{\text{vis},x}-q_{\text{IR},x})\right]\end{aligned} \tag{4.36}$$

$$\begin{aligned}\chi^{(2)}_{zxx,\text{bulk,sep.}} &=\chi^{(2)}_{zyy,\text{bulk,sep.}}\\ &=\frac{1}{(\Delta q_z)_R}\left[\frac{D_{\text{vis}}-D_{\text{IR}}}{2}(q_{\text{vis},z}-q_{\text{IR},z})\right]\end{aligned} \tag{4.37}$$

$$\chi^{(2)}_{zzz,\text{bulk,sep.}} = \frac{1}{(\Delta q_z)_R}\left[\frac{D_{\text{vis}}-D_{\text{IR}}}{2}(q_{\text{vis},z}-q_{\text{IR},z})+\Delta_{\text{vis}}q_{\text{vis},z}+\Delta_{\text{IR}}q_{\text{IR},z}\right] \tag{4.38}$$

while the eight others (Eq. 4.28–4.31) remain unchanged.

4.3 Experimental Methods

Our 10 Hz wavelength-scanning SFG system has been described in Ref. 174 and the phase-sensitive detection and analyis has been illustrated in detail in Refs. 85 and 46. In the present experiments, a collinear configuration of a 532 nm beam (150 μJ per pulse) and mid-IR beam at 2800 cm^{-1} (200 μJ per pulse, 4 cm^{-1} bandwidth) are incident at varying angles with a computer-controlled sample stage and detector arm. The external SFG field (local oscillator, LO) for the heterodyne measurements is generated in transmission from a piece of 50 μm y-cut α-quartz with its optical axis rotated a few degrees from the plane of the incident polarization. Although this provided a low intensity LO, it also minimizes the change in polarization of all beams on account of the birefringence of the y-cut quartz. Data are collected by rotating a 1 mm fused silica phase shifting unit between the LO and sample at each incident angle of interest. The generated signal is a temporal interferogram along the phase-shifting axis and a spectral interferogram along the IR frequency axis.[85] Our phase calibration utilizes a piece of z-cut α-quartz as the reference sample,[175] whose phase has been previously determined in the ssp polarization scheme.[46, 85] Samples consisted of vapour deposited (100 nm) gold over a 5 nm Cr adhesion layer on a glass substrate (EMF, Ithaca NY), cleaned with acetone and anhydrous ethanol prior to measurements. In the ppp polarization scheme, the phase of the z-cut α-quartz is determined by following the derivation of Ref. 176 together with our knowledge of the already determined phase of the z-cut α-quartz in the ssp polarization. We also assume that chiral elements of the nonlinear susceptibility do not contribute significantly to the SFG response of the bulk phase of the quartz, and that surface response is negligible when the plane of incidence is close to the crystallographic axis.[177]

4.4   Results

We introduce the notation that we will use in the remaining sections. $\chi^{(2)}$ originating from the surface dipolar response is abbreviated $\chi^{(2)}_{\text{surf}}$ (first column in Table 4.1). The term $\chi^{(2)}_{\text{comb.surf}}$ is used when describing the sum of all responses independent of the incident angle (first three columns). $\chi^{(2)}_{\text{bulk}}$ is used to refer to the bulk response with incident angle dependence (fourth column). The only exception is the xxz element probed in the ppp polarization scheme, where the $\chi^{(2)}_{xxz,\text{bulk}}$ refers to both the angle-dependent and angle-independent contribution. Any exceptions to this convention will be explicitly stated.

4.4.1   ssp polarization scheme

We have demonstrated that the separable bulk contributions are negligible in the ssp polarization scheme. We drop the subscripts (SFG, vis, IR) from elements of $\mathbf{L}$ and $\mathbf{e}$ as their order and designation is understood. We then arrive at the expression

$$\begin{aligned}
\chi^{(2)}_{\text{ssp}} &= L_{yy}e_y\, L_{yy}e_y\, L_{xx}e_x\, \chi^{(2)}_{yyx,\text{bulk}} + L_{yy}e_y\, L_{yy}e_y\, L_{zz}e_z\, \chi^{(2)}_{yyz,\text{surf}} \\
&\quad + L_{yy}e_y\, L_{yy}e_y\, L_{zz}e_z\, \chi^{(2)}_{yyz,\text{bulk}} \\
&\approx L_{yy}e_y\, L_{yy}e_y\, L_{zz}e_z\, \chi^{(2)}_{yyz,\text{surf}}
\end{aligned} \tag{4.39}$$

where the surface contribution dominates in ssp.

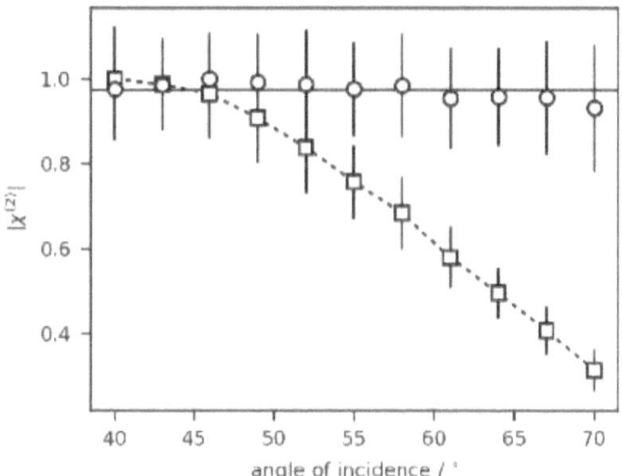

Figure 4.5: Homodyne (intensity) measurements of the magnitude of the effective susceptibility $|\chi_{\text{ssp}}^{(2)}|$ at the air–gold interface as a function of angle (squares). As the data fits well to our model of $|L_{yy}e_y\, L_{yy}e_y\, L_{zz}e_z|$ (dashed line), dividing the data by the local field corrections and unit polarization vectors produces an essentially angle-independent result (circles). Error bars indicate one standard deviation about the mean. The errors are small for large angles due to the higher signal to noise ratio. Reprinted with permission from Ref. 158. Copyright 2020 American Institute of Physics.

Results of homodyne measurements show the magnitude of the effective susceptibility $\chi_{\text{ssp}}^{(2)}$ in Fig. 4.5 (squares) and fit to our model of the local field correction factors (dashed line). Although it was possible to collect gold ssp data at 532 nm with acceptable signal-to-noise, this is not necessarily the case for all metals or at different wavelengths in the visible or near-infrared as a result of the frequency- and angle-dependent local field correction factors, the elements of $\mathbf{L}$ that appear in Eq. 4.39. Upon dividing by these factors, we obtain $\chi_{yyz}^{(2)}$ (circles) that is independent of the angle of incidence (solid line) as expected. This also illustrates our experimental accuracy in maintaining overlap of the beams and alignment into the detector as the sample is rotated by $30°$. The fact that we can remove practically all angle dependent contributions in this treatment also justifies the assumption that $\chi_{yyx,\text{bulk}}^{(2)} \approx \chi_{yyz,\text{bulk}}^{(2)} \approx 0$. We have also measured the phase of $\chi_{yyz,\text{surf}}^{(2)}$ in an ssp heterodyne experiment and determined it to be $65 \pm 2°$ at 2800 cm^{-1}.

4.4.2 ppp polarization scheme

We now turn to our main interest in treating multiple angle of incidence ppp data. Here there are eight elements of $\chi^{(2)}$ that can contribute. In the case of our surface with azimuthal symmetry ($C_{\infty v}$), there are only four non-zero elements that contribute to the surface dipolar response. In the bulk, all eight elements contribute to the overall nonlinear signal. The net result is

$$
\begin{aligned}
\chi^{(2)}_{\text{ppp}} = {} & L_{xx}e_x\,L_{xx}e_x\,L_{zz}e_z\,\chi^{(2)}_{xxz,\text{comb.surf}} + L_{xx}e_x\,L_{zz}e_z\,L_{xx}e_x\,\chi^{(2)}_{xzx,\text{comb.surf}} \\
& + L_{zz}e_z\,L_{xx}e_x\,L_{xx}e_x\,\chi^{(2)}_{zxx,\text{comb.surf}} + L_{zz}e_z\,L_{zz}e_z\,L_{zz}e_z\,\chi^{(2)}_{zzz,\text{comb.surf}} \\
& + L_{xx}e_x\,L_{xx}e_x\,L_{xx}e_x\,\chi^{(2)}_{xxx,\text{bulk}} + L_{xx}e_x\,L_{xx}e_x\,L_{zz}e_z\,\chi^{(2)}_{xxz,\text{bulk}} \\
& + L_{xx}e_x\,L_{zz}e_z z\,L_{xx}e_x\,\chi^{(2)}_{xzx,\text{bulk}} + L_{xx}e_x\,L_{zz}e_z z\,L_{zz}e_z\,\chi^{(2)}_{xzz,\text{bulk}} \\
& + L_{zz}e_z\,L_{xx}e_x\,L_{xx}e_x\,\chi^{(2)}_{zxx,\text{bulk}} + L_{zz}e_z\,L_{xx}e_x\,L_{zz}e_z\,\chi^{(2)}_{zxz,\text{bulk}} \\
& + L_{zz}e_z\,L_{zz}e_z\,L_{xx}e_x\,\chi^{(2)}_{zzx,\text{bulk}} + L_{zz}e_z\,L_{zz}e_z\,L_{zz}e_z\,\chi^{(2)}_{zzz,\text{bulk}}.
\end{aligned}
\tag{4.40}
$$

Some assumptions can be made in order to simplify the above expression. Given the relative size and angle dependence of the $\chi^{(2)}_{\text{bulk}}$ terms, some can be merged (indistinguishable in their angle-dependence) or dropped (negligible magnitude). The model has revealed that all xzz, zzx, and zxz bulk terms are found to be insignificant relative to the rest of the bulk response. We therefore make the approximation that those three terms can be excluded, given that the they do not have corresponding non-zero dipolar terms.

We next examine the angle dependence of the $\chi^{(2)}_{\text{bulk}}$. From the electron gas model, we can see that only the bulk xxx and zxx terms are not constant with respect to the angle of incidence. Thus, we can treat the xxz, xzx and zzz bulk terms as complex-valued constants and merge them with their respective dipolar response, with the exception of the zxx tensor element. According to the model, zxx should not contribute to the surface dipolar response. However, the bulk contribution that is not separable from the overall surface response cannot be ignored. Thus, the overall surface response of the zxx tensor element is not zero, as it originates from the bulk. This term in our expression is merged with its respective bulk term given that we are interested in the extraction of the surface tensor terms. Finally, we arrive at the simplified expression of $\chi_{\mathrm{ppp}}^{(2)}$, where each of the $\chi_{ijk,\mathrm{comb.surf}}^{(2)}$ can be treated as a complex-valued constant

$$
\begin{aligned}
\chi_{\mathrm{ppp}}^{(2)} \approx\ & L_{xx}e_x\, L_{xx}e_x\, L_{zz}e_z\, \chi_{xxz,\mathrm{comb.surf}}^{(2)} + L_{xx}e_x\, L_{zz}e_z\, L_{xx}e_x\, \chi_{xzx,\mathrm{comb.surf}}^{(2)} \\
& + L_{zz}e_z\, L_{zz}e_z\, L_{zz}e_z\, \chi_{zzz,\mathrm{comb.surf}}^{(2)} + L_{xx}e_x\, L_{xx}e_x\, L_{xx}e_x\, \chi_{xxx,\mathrm{bulk}}^{(2)} \\
& + L_{zz}e_z\, L_{xx}e_x\, L_{xx}e_x\, \chi_{zxx,\mathrm{bulk}}^{(2)}.
\end{aligned}
\tag{4.41}
$$

The ppp experiment is performed in the same way as for the ssp polarization scheme, except that we need to take into account that the phase of the LO is shifted by $180°$ upon reflection from the z-cut quartz on either side of the Brewster angle. After the interference fringes have been suitably processed,[46, 85] we arrive at the results shown with points in Fig. 4.6.

To further simplify the expression of $\chi_{\mathrm{ppp}}^{(2)}$, we can represent the two angle dependent bulk elements (zxx and xxx) as well as their respective local field factors as a set of second-order polynomials for their real and imaginary components. This approximation is justified by first examining the linear combination of the zxx and xxx local field factors with the model value of $\chi^{(2)}$ (numbers in parenthesis in Table 4.1). We then fit the line shape for its real and imaginary components with separate second-order polynomial function and found it is reasonable in this approach to capture the line shape in the experimental range of 40–$70°$. Thus, the $\chi_{\mathrm{ppp}}^{(2)}$ expression for the multi-angle of incidence fitting becomes

$$
\begin{aligned}
\chi_{\mathrm{ppp}}^{(2)} \approx\ & (LLLeee)_{xxz}(\theta)\, \chi_{xxz,\mathrm{surf+bulk}}^{(2)} + (LLLeee)_{xxz}(\theta)\, \chi_{xzx,\mathrm{surf+bulk}}^{(2)} \\
& + (LLLeee)_{zzz}(\theta)\, \chi_{zzz,\mathrm{surf+bulk}}^{(2)} + a\theta^2 + b\theta + c,
\end{aligned}
\tag{4.42}
$$

where $(LLLeee)_{ijk}$ is the product of the local field factors and unit polarization vectors for all three fields (SFG, visible, IR) for a given tensor elements. The coefficients a, b and c encompass the linear combination of the product of the local field factors, unit polarization vectors, and $\chi_{\mathrm{bulk}}^{(2)}$ for the zxx and xxx components. As a result of the angle dependence of $\mathbf{L}$, (see Fig. 4.4) and the two angle dependent $\chi^{(2)}$ terms (zxx and xxx), we can solve for all the surface tensor elements. From our fitting, we find that $|xzx/zzz| = 0.39 \pm 0.02$, $|xxz/zzz| = 1.11 \pm 0.06$, $\phi_{xzx} = -146 \pm 3°$, $\phi_{xxz} = 62 \pm 3°$, and $\phi_{zzz} = 83 \pm 3°$.

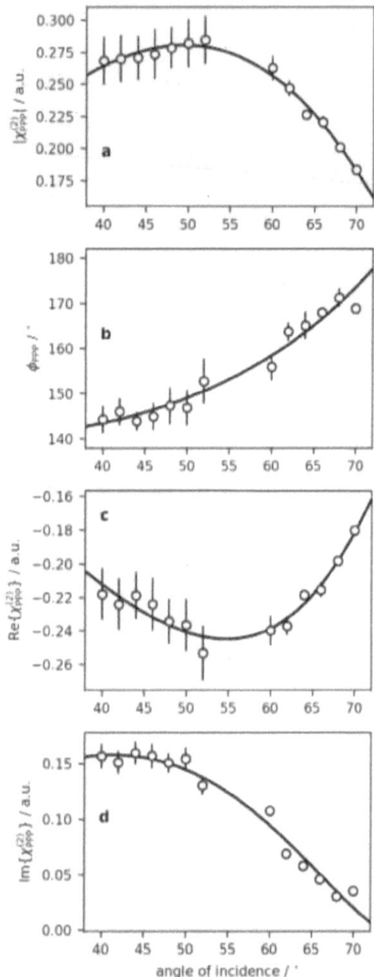

Figure 4.6: Experimental results (points) from the heterodyne measurements in the ppp polarization scheme displaying the (a) magnitude, (b) phase, (c) real, and (d) imaginary components of $\chi^{(2)}_{\mathrm{ppp}}$. A fit of this data to Eq. 4.42 is shown with the lines in each panel. Error bars indicate one standard deviation about the mean. Reprinted with permission from Ref. 158. Copyright 2020 American Institute of Physics.

4.5 Discussion

Before bench-marking our experimental fitting results, it is important to first understand what each of the simplified tensor elements represents; the three elements denoted as 'surf+bulk' are especially interesting. In the case of the xzx component, the bulk response is significant relative to the overall bulk contribution. Thus, the xzx component is the combination of the dipolar surface contribution as well as the quadrupolar bulk contribution. For the case of the xxz and zzz elements, their quadrupolar bulk contribution are insignificant relative to the rest. Thus, as an approximation, we can treat the two tensor elements as the dipolar surface response only, where $\chi^{(2)}_{ijk,\text{comb.surf}} \approx \chi^{(2)}_{ijk,\text{surf}}$.

There are two tests we can perform in order to validate our results. As our gold surface has $C_{\infty v}$ symmetry, we know that $\chi^{(2)}_{xxz,\text{surf}} = \chi^{(2)}_{yyz,\text{surf}}$ for both the magnitude and phase of these tensor elements. Our first check is therefore to compare the result $\phi_{xxz,\text{surf}} = 62 \pm 3°$ obtained in our deconstruction of the ppp data to our result of $\phi_{yyz,\text{surf}} = 65 \pm 2°$ measured directly in the ssp heterodyne experiment; we find them to be in agreement. The good match also validates our model result that the relative contribution of $\chi^{(2)}_{xxz,\text{bulk}}$ is predicted to be small. The second check is a comparison between experimental ppp/ssp ratio (points in Fig. 4.7) and the predictions of our model in extracting the magnitude of all tensor elements contributing to the ppp signal. If we take the magnitude of $\chi^{(2)}_{xxz}$ from the ppp fit and multiply by the magnitude of L_{yyz} we should reproduce the ssp data according to

$$
\begin{aligned}
|\chi^{(2)}_{\text{ssp}}| &\approx |L_{yy}e_y\, L_{yy}e_y\, L_{zz}e_z\, \chi^{(2)}_{yyz,\text{surf}}| \\
&= |L_{yy}e_y\, L_{yy}e_y\, L_{zz}e_z\, \chi^{(2)}_{xxz,\text{surf}}|.
\end{aligned}
\tag{4.43}
$$

The predicted ratio is plotted with the solid line in Fig. 4.7. The excellent agreement further validates the decomposition of $\chi^{(2)}$ tensor elements with our multi-angle ppp experiment. Note that, although incorporation of the low-intensity ssp data has introduced large error bars at high angles, this is not a fit (no adjustable parameters), merely a comparison of two independent results.

The decomposition of the three surface $\chi^{(2)}$ elements with their respective local field

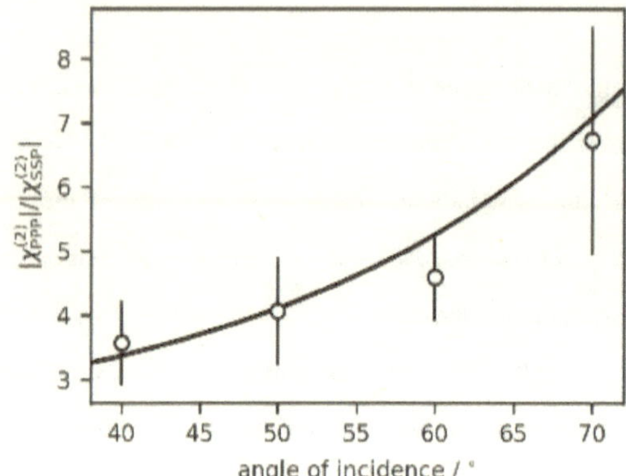

Figure 4.7: Comparison of the ratio of the magnitudes of the effective susceptibilities measured in separate ppp and ssp homodyne experiments (points), to that calculated from the ppp multi-angle heterodyne data (line), demonstrating the equivalence expressed in Eq. 4.43. Error bars represent one standard deviation about the mean. Reprinted with permission from Ref. 158. Copyright 2020 American Institute of Physics.

factors and the total bulk contribution is presented in Fig. 4.8. Without accounting for local field effects, the relative size of the three surface tensor elements follows the order of $xxz >$ $zzz > xzx$. In comparison, it is found that the experimental gold $\chi_{\mathrm{ppp}}^{(2)}$ response is dominated by the effective surface xzx term followed by the xxz term then the zzz term. This is due to the local field factor essentially acts as a relative weighting factor to its respective tensor element. More interestingly, the total contribution of the bulk response is significantly smaller than the rest of the surface response in Fig. 4.8. Due to the screening of the z-component inside the gold when crossing the interface, components parallel to the surface are much larger than the perpendicular ones. This result is in agreement with Refs. 47 and 178, given that the surface $\chi^{(2)}$ elements contain both the isotropic and inseparable bulk contributions.

From the model result in Table 4.1, there should be a significant contribution from the xxx and zxx elements relative to the total bulk response. Furthermore, given the relative size of the xxx and zxx local field factors, one would expect that the total bulk contribution

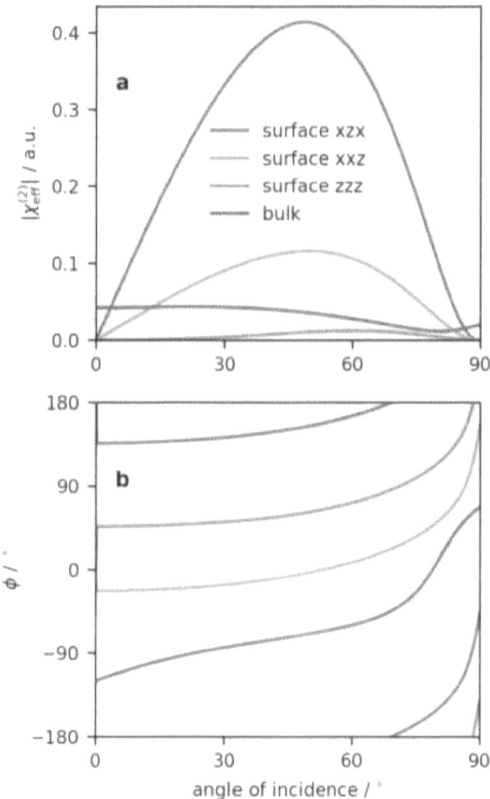

Figure 4.8: Comparison of the (a) magnitude and (b) phase of the product of the susceptibility from the fitting and its respective local field factor. Reprinted with permission from Ref. 158. Copyright 2020 American Institute of Physics.

should be significant compared to the surface response. We conclude that the total separable bulk contribution is smaller than the surface and inseparable bulk response. This is in agreement with what has been reported for other metals, in studies where oxidation or disordering decreases the nonlinear response.[172]

4.6   Conclusions

A heterodyne-detected multi-angle approach has been demonstrated to decompose the second-order nonlinear susceptibility measured in an SFG experiment in order to obtain the substrate optical properties. As a demonstration, we have performed a study of a planar gold surface that necessitates consideration of dipolar surface contributions as well as quadrupolar bulk contributions. This method is especially interesting and relevant in cases where only ppp experiments produce appreciable signals on account of surface selection rules. In the case where organic molecules are adsorbed on metal surfaces, this method has the potential to assist in separating substrate and adsorbate components of the nonlinear susceptibility tensor. While additional challenges remain in the case of vibrationally-resonant contributions, the work we have presented here provide the necessary theoretical framework, experimental techniques and results to embark on such studies.

Chapter 5

Temperature-Dependent Structure of an Alkyl Phosphate Surfactant on Iron

5.1 Introduction

Surfactants have been the front-line prevention in the field of corrosion inhibition for many decades.[179–183] Alkyl phosphate surfactants have shown to be promising in combating corrosion on steel and iron surfaces that have been subsequently deployed in many infrastructures and machinery such as pipelines and heat-exchangers. Given that iron is such a widely used material, searching for alternative corrosion inhibitors that are low cost and low toxicity has been a subject of great interest.[184–188] The importance of understanding how these corrosion inhibitors are structured and function on metal surfaces is crucial in designing and engineering more efficient and effective surfactants. Even-though the use of alkyl phosphates for corrosion inhibition has been ongoing, a fundamental understanding of how these molecules structure on metal surfaces and its potential structural changes at elevated temperature are not well understood.[189–191] In this work, we present a nonlinear optical study of an alkyl phosphate surfactant, bis-2-ethylhexyl phosphate (BEP) adsorbed on iron. We also examine the possible structural changes of BEP at elevated temperature.

Vibrational sum frequency generation spectroscopy has been utilized in the study of organic-metal interface for quite some time.[32,43,45,47–51,87,192–195] Due to the complexity

Element	wt% w.r.t. C	atom % w.r.t. C
C	36.12	64.71
Fe	52.53	20.24
O	11.06	14.87
Ca	0.18	0.10
Al	0.06	0.05
Cu	0.05	0.02
Si	0.02	0.01

Table 5.1: Tabulation of the elemental analysis of the iron sample from EDX measurements. The carbon content came from the graphite paste used underneath the sputtered iron.

of both the local field effect and the non-resonant contribution from the metal, the quantitative study of an organic molecule structured on a metal surface still poses a significant challenge. In addition, many SFG metal interface studies face the issue of the inaccessibility of polarization geometry due to the surface selection rule of the metal.[115] Therefore in this work, we demonstrate our multi-angle of incidence approach from Chapter 4 with a combination of homodyne and heterodyne SFG to study the BEP-Fe interface in air by extracting all contributing second order nonlinear susceptibility tensor elements when all beams are p-polarized. Furthermore, we examine the effect of temperature on the BEP by comparing the methyl function group orientation between the room temperature sample and a heat-treated sample.

5.2 Methods

5.2.1 Sample preparation

The iron substrate is made from sputtering 100 nm of iron onto a 25 mm $\times$ 75 mm microscope slide that was cleaned using isopropanol sonication for 60 min. The vapour deposition is performed in a deposition system (Mantis deposition Ltd.) with 99.9% iron target (Kurt J. Lesker Company) with a pump pressure of approximately 10^{-3} mbar. The iron thickness is determined by the built-in quartz crystal microbalance (QCM) as well as

Figure 5.1: SEM image of an iron substrate at 30000 times magnification with an accelerating voltage of 13.0 kV.

a separate profilometry measurement (Bruker Dektak). The elemental composition of the iron film is determined by energy dispersive X-Ray (Bruker Quantax) analysis on a field emission scanning electron microscope (Hitachi S-4800) coupled with the EDX System. The measurement was performed in triplicate with a modified iron film that is sputtered onto a graphite covered microscope slide to prevent the glass substrate from altering the ratio of the iron and oxygen atoms. The elemental composition for the sputtered iron sample is shown in Table 5.1 and the surface of the substrate is shown in Figure 5.1

Bis(2-ethylhexyl) phosphate (BEP, Sigma-aldrich) was diluted in 99.97% spectral grade heptane (Sigma-aldrich) to make a 0.1% w/w solution. The formation of the surfactant layer onto the iron substrate is achieved by submerging the iron into the solution for 60 min then rinsing with neat heptane and air drying. The sample was subsequently left for another 24 h covered with a beaker before any spectroscopic measurements. The heat-treated sample was annealed using a custom made heating element and the surface temperature was recorded to ensure that it reaches 200 °C.

5.2.2 Sum frequency generation spectroscopy

Our 10 Hz tunable wavelength scanning SFG system has been described in Ref. 174 and the basic heterodyne measurement procedure has been illustrated in detail in Refs. 85 and 46. The incident angle-resolved heterodyne basics have been described in Chapter 4 and Ref. 158. Here, a p-polarized IR beam (150 μJ/pulse) is spatially and temporally overlapped with a p-polarized visible beam at 532 nm (60 μJ/pulse, around one third of the typical output) in a non-collinear geometry for the homodyne measurement. We have observed that higher energy (over 60 μJ/pulse from the visible beam) results in the damage of the iron film. In the non-collinear geometry, the IR beam is 7° greater in the incident angle than the visible beam. The homodyne measurements are taken with 100 laser shots average for all frequency across all incident angle spectra. The heterodyne phase measurement are taken with 50 laser shots average for each PSU tilt angle step for both sample and reference z-cut quartz in the collinear geometry.

5.2.3 Spectroscopic ellipsometry

5.2.3.1 Visible wavelengths

The complex refractive index of the sputtered iron film in the visible region was obtained from a commercial spectroscopic ellipsometer (J.A. Woollam alpha-SE). The spectra of Ψ and Δ are taken at 65°, 70°, and 75° incident angles with three replicates each. A two semi-infinite layer (air-metal) model is considered at both temperatures (r.t. and 200 °) given the thickness of the sputtered iron is 100 nm and the transmittance at these wavelengths is on the order of 10^{-4}.

5.2.3.2 Mid-infrared wavelengths

The complex refractive index of the iron substrate in the mid infrared region is obtained from a custom rotating compensator ellipsometer (RCE) that we constructed for this purpose. In our PCSA setup as shown in Figure 5.2, a Bruker Vertex 70 FTIR is used

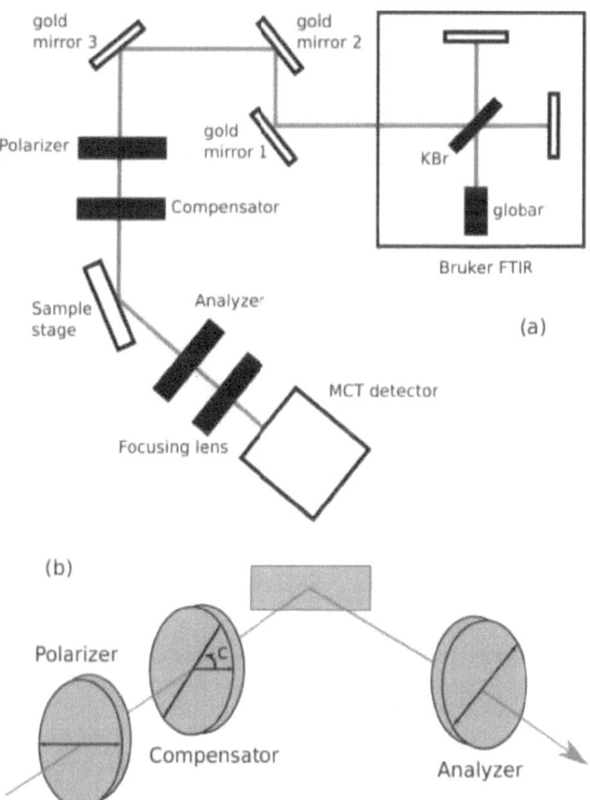

Figure 5.2: (a) The experimental configuration of the spectroscopic mid-IR ellipsometer. The gold mirror 3 in the figure is a gold coated concaved mirror with a focal length of 50.0 mm where the compensator is placed within. (b) The general scheme of the PCSA ellipsometer illustrating the azimuthal angles of the polarization optics, where the compensator angle is denoted C.

as the mid-IR source, while the polarizer and the analyzer are both holographic wire grid polarizer on zinc selenide (ZeSe) substrate (Thorlabs, WP25H-Z). The compensator used in the setup is a 3000 nm zero-order $\lambda/4$ waveplate (Altechna, 2-IRPW-ZO-L/4-3000-C) and the detector is a 1 mm^2 mercury cadmium telluride (MCT) detector with fast pre-amplifier (Kolmar Technologies). A 50 mm ZeSe lens (Thorlabs, LA7656-G) is used to focus the light into the entrance slit of the detector. The rationale of choosing an RCE setup is its advantage over a rotating analyzer ellipsometer due to the unambiguous determination of the ellipsometric angle, Δ. In short, the measurable quantity from the rotating analyzer ellipsometer is $\cos\Delta$. It therefore cannot determine the sign of Δ. In addition, the Δ region close to $0°$ or $180°$ is difficult to obtain for a rotating analyzer setup due to the inversion of the cosine function. Therefore, we have adopted a polarizer-compensator-sample-analyzer (PCSA) setup based on a modification to the instrument described in Ref 196.

By evaluating the first component of the Stokes vector describing the PCSA setup, the intensity has harmonic contributions of the second and the fourth order in the compensator azimuth angle, C. Therefore, we can write the measurable signal as[196]

$$I(C) = A_0 + A_2 \cos 2C + B_2 \sin 2C + A_4 \cos 4C + B_4 \sin 4C, \tag{5.1}$$

where each of the Fourier coefficients can be described as

$$
\begin{aligned}
A_0 &= \frac{1}{\pi} \int_0^\pi I(C)\, dC \\
A_2 &= \frac{2}{\pi} \int_0^\pi I(C) \cos 2C\, dC \\
B_2 &= \frac{2}{\pi} \int_0^\pi I(C) \sin 2C\, dC \\
A_4 &= \frac{2}{\pi} \int_0^\pi I(C) \cos 4C\, dC \\
B_4 &= \frac{2}{\pi} \int_0^\pi I(C) \sin 4C\, dC.
\end{aligned}
\tag{5.2}
$$

For the PCSA ellipsometer, each of the Fourier coefficients can be related to experimental Ψ and Δ by

$$A_0 = 1 - \cos 2A \cos 2\Psi + \frac{1}{2}(1 + y_c)[\cos 2P(\cos 2A - \cos 2\Psi) + \sin 2P \sin 2A \sin 2\Psi \cos \Delta]$$

$$A_2 = x_c(\cos 2A - \cos 2\Psi) + x_c \cos 2P(1 - \cos 2A \cos 2\Psi) - z_c \sin 2P \sin 2A \sin 2\Psi \sin \Delta$$

$$B_2 = x_c \sin 2A \sin 2\Psi \cos \Delta + z_c \cos 2P \sin 2A \sin 2\Psi \sin \Delta + x_c \sin 2P(1 - \cos 2A \cos 2\Psi)$$

$$A_4 = \frac{1}{2}(1 - y_c)[\cos 2P(\cos 2A - \cos 2\Psi) - \sin 2P \sin 2A \sin 2\Psi \cos \Delta]$$

$$B_4 = \frac{1}{2}(1 - y_c)[\cos 2P \sin 2A \sin 2\Psi \cos \Delta + \sin 2P(\cos 2A - \cos 2\Psi)],$$

$$(5.3)$$

where the x_c, y_c, and z_c are the non-ideal parameters for the compensators and the P and A are the azimuthal angle of the polarizer and analyzer, respectively. By rearranging Eq. 5.3, we can obtain

$$\Psi = \arctan\left(\sqrt{\left(\frac{x_c}{z_c}\frac{B_4}{A_4} - \frac{1 - y_c}{2z_c}\frac{B_2}{A_4}\right)^2 + \left(\frac{-B_4}{A_4}\right)^2}\right)/2$$

$$\Delta = \arctan\left(\left(\frac{x_c}{z_c}\frac{B_4}{A_4} - \frac{1 - y_c}{2z_c}\frac{B_2}{A_4}\right)/\left(\frac{-B_4}{A_4}\right)\right).$$

$$(5.4)$$

Since both Ψ and Δ are calculated from the root of the arctangent function (which only returns -90° to 90°), therefore the exact degree of angle has to be derived from the sign of the Fourier coefficients as

$$A_4 > 0 \longrightarrow \Psi = 90° - \Psi$$

$$A_4 < 0 \longrightarrow \Psi = \Psi$$

$$B_4 > 0 \longrightarrow \Delta = \Delta$$

$$B_4 < 0 \longrightarrow \Delta = 180° + \Delta.$$

$$(5.5)$$

Finally, the non-ideal parameters for the compensator can be described as

$$x_c = \cos 2\Psi_c$$

$$y_c = \sin 2\Psi_c \cos \Delta_c$$

$$z_c = \sin 2\Psi_c \sin \Delta_c$$

$$(5.6)$$

where $\tan \Psi_c$ and Δ_c are the relative amplitude change between the fast/slow axis and the retardation of the compensator, respectively. Both Ψ_c and Δ_c can be obtained experimentally in a straight through geometry in transmission mode with the sample stage.

In the straight through configuration, the Fourier coefficients can be obtained as

$$A_0 = 1$$

$$A_2 = x_c(\cos 2C + \sin 2C)$$

$$B_2 = x_c(-\sin 2C + \cos 2C)$$

$$A_4 = \frac{1}{2}(1 - y_c)\sin 4C$$

$$B_4 = \frac{1}{2}(1 - y_c)\cos 4C, \tag{5.7}$$

where C here refers to the azimuthal angle of the compensator relative to the polarizer. The non-ideal parameters of the compensator can be calculated from Eq. 5.7 as

$$x_c = \pm \frac{\sqrt{2A_2^2 + 2B_2^2}}{2A_0}$$

$$y_c = 1 - 2\frac{\sqrt{A_4^2 + B_4^2}}{A_0} \tag{5.8}$$

and z_c can then be calculated from Eq. 5.6 in combination with Eq. 5.8. Lastly, additional calibration regarding to the error between the position of the fast axis and the rotation motor readout also needs to be considered. In the straight through configuration, by crossing the polarizer and the analyzer, the fourth harmonic Fourier coefficients can be expressed as

$$A_4 = -\frac{1}{2}(1 - y_c)(1 - \cos 2\Psi)\cos 4C'$$

$$B_4 = \frac{1}{2}(1 - y_c)(1 - \cos 2\Psi)\sin 4C', \tag{5.9}$$

where $C' = \arctan(-B_4/A_4)$ is the degree of error associated with the experimental position of the compensator.

For the calibration, a low frequency (345 Hz) chopper is placed in the position after the compensator that is connected to a lock-in amplifier (Stanford Research Systems SR830). The absolute azimuth of polarizer is first determined by removing all optics except for the polarizer in the reflection geometry. The Brewster's angle of the fused silica at 3.33 μm is used as the incident angle. The signal is then minimized while computer controlled motor is stepped in 0.1° per step with a piece of 8 mm fused silica on the sample stage. Once calibrated, the rest of the optics' azimuthal angle is calibrated with respect to the polarizer. The analyzer is then placed back into the optical path in straight through geometry to maximize the signal to determine the $0°$ position of the analyzer relative to the polarizer. Finally the compensator can be place back into the optical path to determine the relative azimuthal angle in the same manner as the analyzer. The calibration of the position error of the compensator and the determination of Ψ_c and Δ_c is done in a fashion described above. All spectra of the samples are taken at $60°$ incident angle in the reflected geometry while stepping the compensator at $5°$ per step. The spectra collection in the frequency range of 2000 cm^{-1} to 4000 cm^{-1} with 64 averages in triplicates.

5.2.4 Electrochemical impedance spectroscopy

A custom Teflon eletrochemical cell was used where the sample can be inserted from the bottom as a working electrode and the contact area is 10 mm in diameter. A 25 mm $\times$ 25 mm 99.9% platinum foil (Sigma-Aldrich, 267244-350MG) was spot welded to the platinum wire as the counter electrode. A custom reversible hydrogen reference electrode was used. The electrochemical measurements were performed with a potentiostat (Gamry Instruments, Reference 600) at open circuit potential from an initial frequency of 60 kHz to final frequency of 20 mHz. The electrolyte used in the experiment is 0.1 M spectral grade sulfuric acid (Spectrum Chemical 7664-93-9) in 18.2 M$\Omega \cdot$cm (Milli-Q) water. Experimental data was analyzed and fitted using ZView.

5.3 Results & Discussion

5.3.1 Electrochemical impedance measurements

In order to quantitatively understand the effect of temperature with respect to the efficiency of the corrosion inhibition ability of BEP on iron, we have opted to measure the charge-transfer resistance (R_{ct}) via electrochemical impedance spectroscopy. By comparing the room temperature sample with the 200 °C heat-treated sample, we can have a quantitative

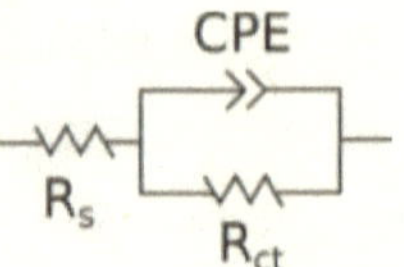

Figure 5.3: Equivalent circuit used for the fitting of the experimental electrochemical system

comparison of their relative resistance for the electrochemical process of corrosion. We have chosen to fit our experimental impedance data to a simplified Randles circuit following Ref. 197 as shown in Fig. 5.3. Although such an equivalent circuit may not accurately account for the complex electrochemical system with surfactant structured on iron,[198] it provides a straightforward route to compare the relative difference between the room temperature sample to the heat-treated one. We have opted to use a constant phase element (CPE) in the Randles circuit given that the capacitance dispersion effect in the electric double layer region is well known.[199,200] Such effect is often attributed to the inhomogeneities of the electrode surface.[199–201] The constant phase element can be described as[202]

$$Z_{CPE} = \frac{1}{A(j\omega)^\alpha} \tag{5.10}$$

where A is the constant phase element magnitude in units of $\Omega^{-1} \cdot cm^{-2} \cdot s^\alpha$ and the exponent α is an indication of the behaviour of the CPE to be either capacitor-like in nature (α close to 1) or resistor-like nature (α close to 0).[203,204] In other words, a CPE is merely a mathematical construct that mimics the behaviour of many electrochemical interfaces. The equivalent capacitance can be determined from

$$C_{eq} = (AR^{(1-\alpha)})^{\frac{1}{\alpha}} \tag{5.11}$$

where R here can be approximated by the solution resistance, R_s when $R_{ct} \gg R_s$.[205] Both A and α can be determined through the fitting of the electrochemical impedance data.[202] By adopting the simplified Randles equivalent circuit, we can then determine the resistance from the electrolyte (R_s), the characteristics of the CPE, and the resistance associated with

sample	blank Fe	r.t. BEP-Fe	heat-treated BEP-Fe
R_s / $\Omega\cdot\text{cm}^2$	181.0(1.0)	153.4(1.7)	168.9(0.8)
A / $\text{mF}\cdot\text{cm}^{-2}$	1.181(0.011)	0.177(0.003)	0.425(0.003)
α / $\text{F}\cdot\text{cm}^{-2}\cdot\text{s}^{(\alpha-1)}$	0.911(0.006)	0.813(0.007)	0.907(0.004)
R_{ct} / $\Omega\cdot\text{cm}^2$	14165(880)	29741(2656)	18241(691)
E_{oc} / V	0.713	0.703	0.718

Table 5.2: Electrochemical impedance measurement fitting result obtained from the complex fitting approach with calc-modulus data weighting. Values in round brackets indicate fitting errors.

the charge transfer at the electrode–electrolyte interface (R_{ct}). Given that the corrosion current density is inversely proportional to the charge transfer resistance given by the Stern-Geary relationship,[206,207] we can quantitatively compare the inhibition effectiveness of BEP on iron as well as the effect of temperature on its effectiveness as a corrosion inhibitor.

The trend of R_{ct} obtained from the fitting of the electrochemical impedance measurement is in agreement with literature results that R_{ct} increases with the adsorption of the BEP on iron surface.[190] In addition, the effectiveness of the BEP as a corrosion inhibitor decreases in elevated temperature. Thus, R_{ct} from the heat-treated sample is significantly lower than that of the room temperature sample.

Upon examination of the experimental result of the measurement, the measured impedance is significantly higher than that of similar experiments.[197,208] The fitting results have shown the R_{ct} to be in the kΩs range and the solution resistance, R_s to be in the range of 100 Ω for a 0.1 M H_2SO_4 solution as shown in Figure 5.4 and Figure 5.5. The measurement has also been repeated several times in order to validate its reproducibility. The high solution resistance is likely attributed by the poor contact between the external lead and the working electrode where the contact resistance contributes to the overall measured resistance.[209] The disagreement of the charge transfer resistance between our experimental result and the literature value can be explained by the difference in the iron substrate. In our system, the working electrode is a 100 nm iron oxide film that is deposited onto a piece

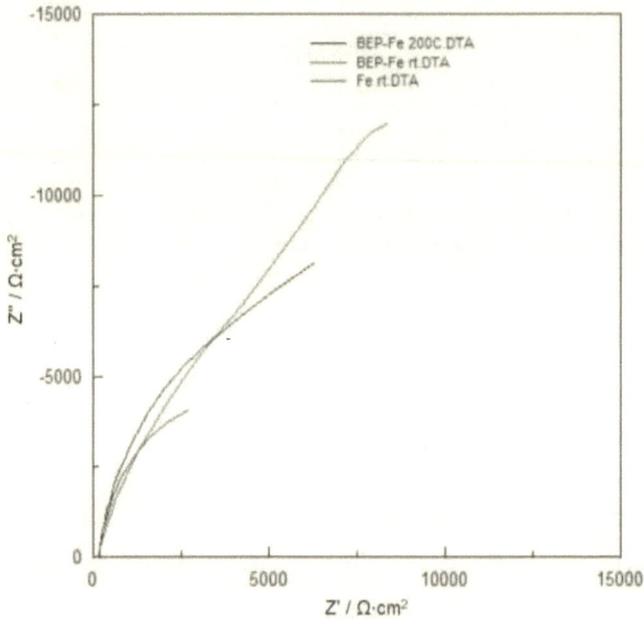

Figure 5.4: Nyquist plot of the experimental impedance for all blank Fe sample, room temperature BEP-Fe sample, and 200 °C heat-treated BE-Fe sample.

of microscope slide, whereas a piece of bulk iron electrode is used for the measurement in Ref. 197. Ref. 210 has reported similar charge transfer impedance for a thin iron oxide film made from physical vapour deposition which agrees with our result. Given that all trials are all taken with the same experimental setup and the resulting R_{ct} is also in agreement with our expectation as shown in Table 5.2, we have preliminarily determined that a 60% reduction in the corrosion inhibition efficiency of the BEP is observed at 200° temperature.

5.3.2   Mid-IR spectroscopic ellipsometer calibration

We first present the characterization of the zero-order $\lambda/4$ waveplate as the compensator in the PCSA setup as shown in Fig. 5.6. It can be seem from the expression of the non-

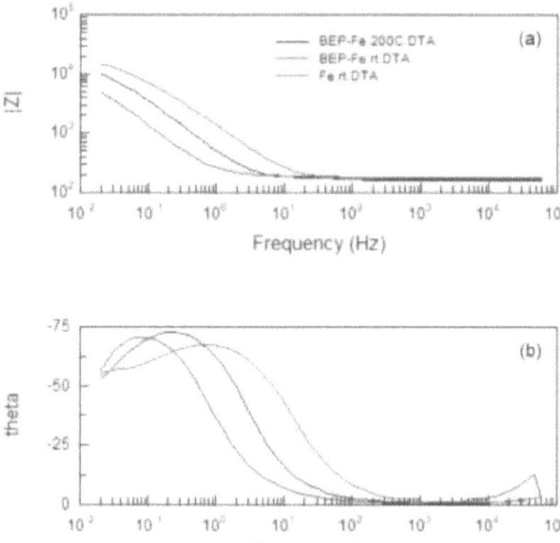

Figure 5.5: Experimental magnitude (a) and phase (b) of the impedance for all blank Fe sample, room temperature BEP-Fe sample, and 200 °C heat-treated BE-Fe sample.

ideal compensator parameters in Eq. 5.6 that when the retardation is close or equal to $0°$ or $180°$, the z_c term becomes zero. Additionally, the $\cos \Delta_c$ term in y_c approaches unity. This results in the inability to reliably determine the Fourier coefficients accurately. In the frequency range of interest between 2800 cm^{-1} and 3100 cm^{-1}, the experimentally determined retardation of the waveplate is centred around $70°$, which is a $20°$ deviation from the ideal retardation of $90°$. Furthermore, given that the retardation of a $\lambda/4$ waveplate is dependent upon it frequency, the non-ideal compensator term will need to be calculated at each frequency collected for the subsequent measurements.

We have bench-marked our PCSA rotating compensator ellipsometer using two reference substrates, fused silica and zinc selenide. Both measurements are performed at $60°$ incident angle. Figure 5.7 shows the experimentally determined refractive index of a piece of IR grade fused silica as well as the $\tan \Psi$ and Δ as a function of frequency. The

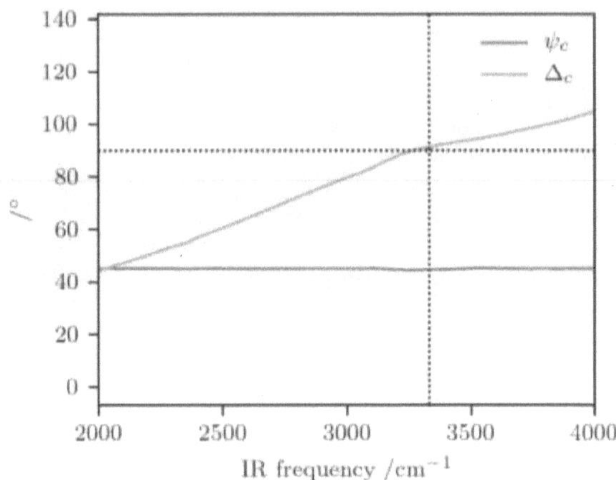

Figure 5.6: Retardation (Δ_c) and relative amplitude change (Ψ_c) of the compensator, where the intercept of the dashed line indicates the specification of the retardation provided by the manufacturer.

artifact shown in the Δ plot stems from the lineshape contribution of the lamp profile from the FTIR where imperfect division occurred during background correction. In addition, the CO_2 peak at 2341 cm^{-1} is also present in the Δ plot attributed by the imperfect background correction. The literature result of the refractive index is obtained from Ref. 211. Further comparison of the literature result to our experimental determination of the refractive index for zinc selenide as well is shown in Figure 5.8 where the literature optical constant is obtained from Ref. 212.

5.3.3 Refractive index of iron substrates

Five literature refractive indices of iron in the visible wavelength region are plotted in Figure 5.9.[213–215] The degree of deviation between each literature result can be attributed to the composition of the iron, as well as the method of which it is obtained. Therefore, without further knowledge of how these values are determined, one should not simply choose to apply these literature refractive indices in order to calculate the local field

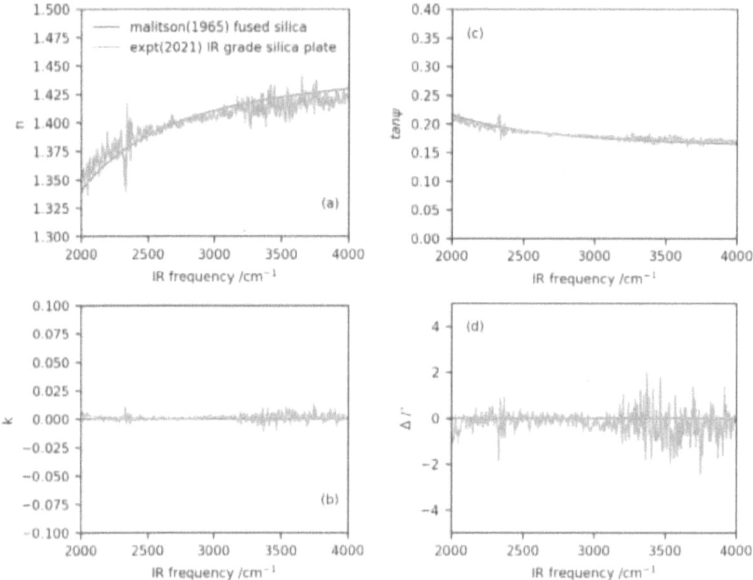

Figure 5.7: Panel (a), (b) show the real and imaginary component of the refractive index of fused silica and panel (c), (d) show the $|\rho|$ or $\tan\Psi$ and Δ of fused silica at an incident angle of $60°$.

correction factors for quantitative analysis of SFG data.

The experimental Ψ and Δ measurements in the visible wavelength region are collected on the alpha-SE with three incident angles in triplicates as shown in Figure 5.10. A two semi-infinite layer model was applied by calculating the resulting refractive indices using the wavelength-by-wavelength approach to determine the complex refractive index of iron substrate for both room temperature and 200 °C heat-treated sample in the visible wavelength region as shown in Figure 5.11. We note that there is a significant change in the refractive index upon heat-treatment in both the absorption and dispersion lineshape. The experimentally determined Ψ and Δ of the iron substrate in the mid-IR region for both room temperature and the heat-treated sample is plotted in Figure 5.12 along with the calculated complex refractive indices.

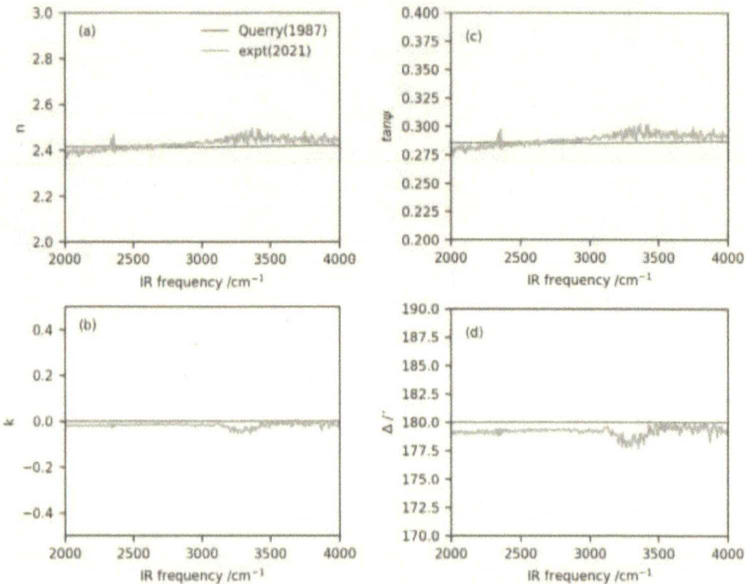

Figure 5.8: Panel (a), (b) show the real and imaginary component of the refractive index of zinc selenide and panel (c), (d) show the $|\rho|$ or $\tan\Psi$ and Δ of zinc selenide at an incident angle of 60°.

5.3.4 Local field correction factors

We first present the local field effect correction factors for the room temperature sample as a function of the AOI for both the non-collinear and the collinear geometry for our setup where in the non-collinear geometry, the IR beam is 7° greater than that of the visible beam. Since the homodyne measurements are performed in the non-collinear geometry while the phase measurement can only be perform in a collinear geometry in our current setup, we first need to resolve any potential discrepancy between their local field factors. We first examine the local fields of the two geometries as shown in Figure 5.13 at 2800 cm^{-1} in IR frequency. Upon closer examination, we can conclude that both collinear and non-collinear geometry measurements should yield identical $\chi^{(2)}_{\text{eff}}$ spectra as a function of the

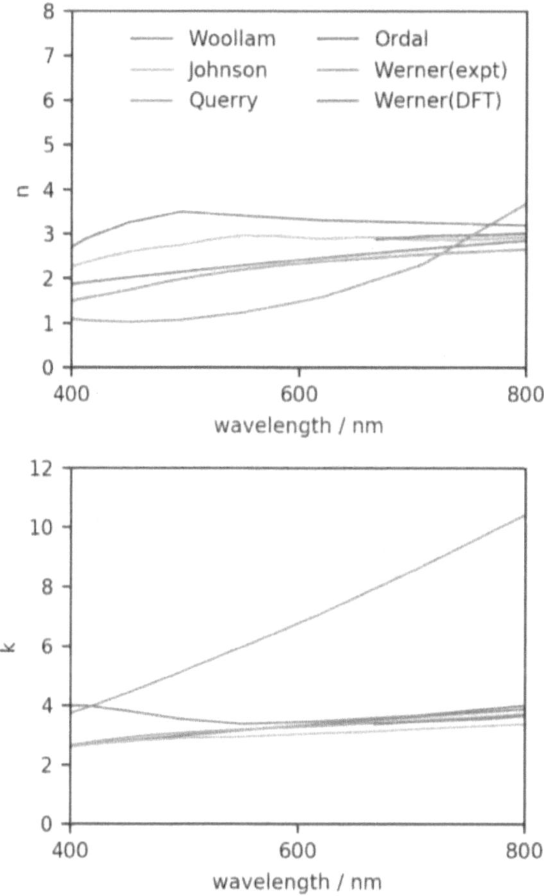

Figure 5.9: Complex refractive indices of iron from the literature in the visible region.[213–215]

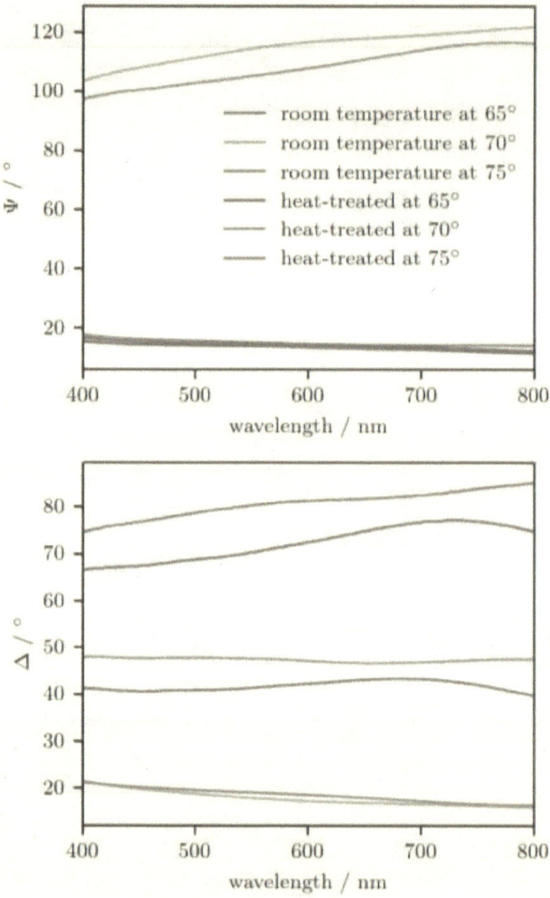

Figure 5.10: Experimental Ψ and Δ of the iron substrate obtained using the wavelength-by-wavelength approach.

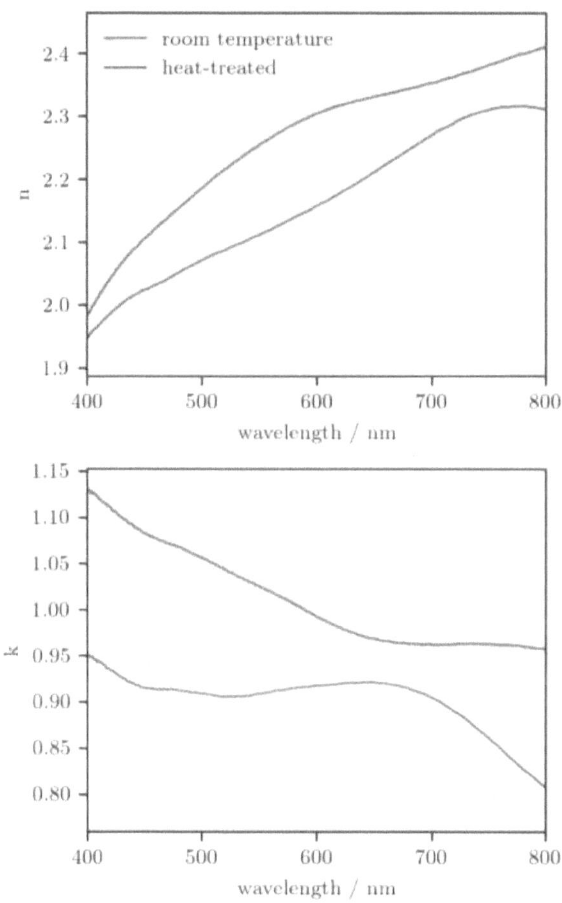

Figure 5.11: Experimental complex refractive index of iron substrate calculated using the experimental Ψ and Δ with a two layer semi-infinite model.

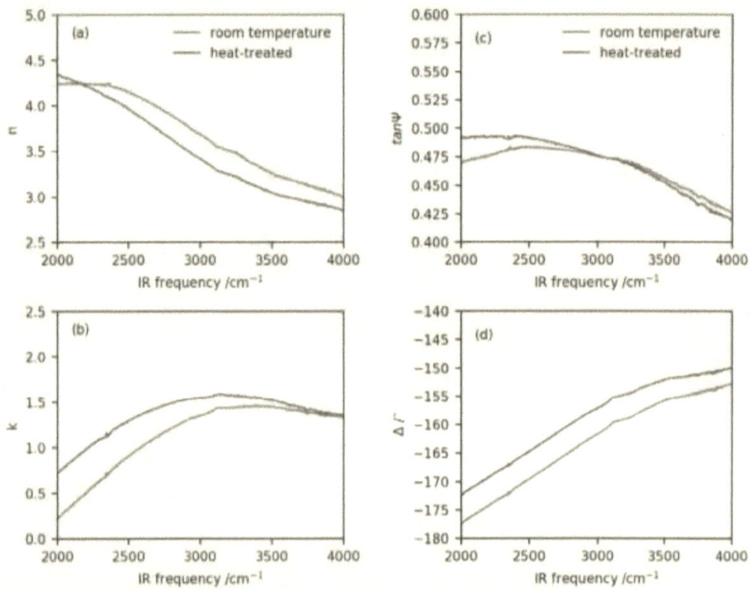

Figure 5.12: Panel (a), (b) show the real and imaginary component of the refractive index and panel (c), (d) show the $|\rho|$ or $\tan\Psi$ and Δ at an incident angle of $60°$.

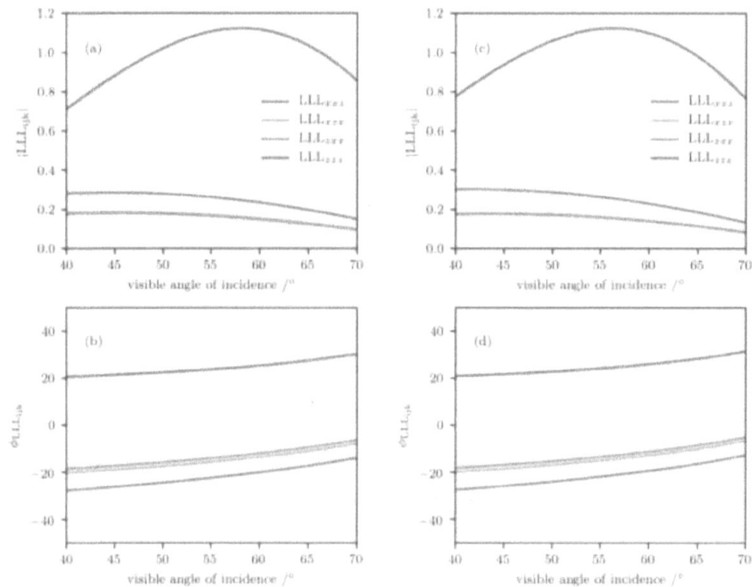

Figure 5.13: (a), (b) are the magnitude and the phase of the local field correction factor of the collinear geometry for the room temperature sample, respectively. (c), (d) are the magnitude and the phase of the local field correction factor of the non-collinear geometry for the room temperature sample, respectively.

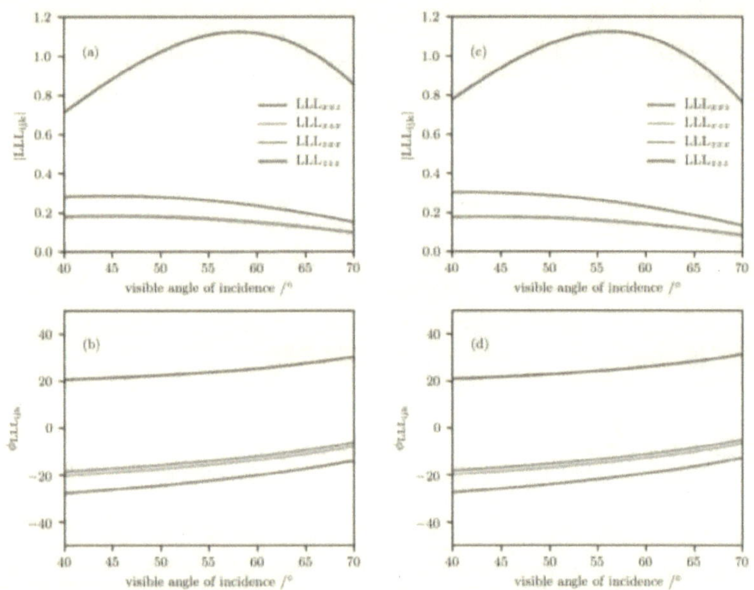

Figure 5.14: (a), (b) are the magnitude and the phase of the local field correction factor of the collinear geometry for the 200 °C heat treated sample, respectively. (c), (d) are the magnitude and the phase of the local field correction factor of the non-collinear geometry for the 200 °C heat treated sample, respectively.

AOI, except at higher AOI in our measurement range where the magnitude of $L_{xx}L_{xx}L_{zz}$, $L_{xx}L_{zz}L_{xx}$, and $L_{zz}L_{xx}L_{xx}$ deviate between the two geometries in a minor way. However, in the best case scenario, both homodyne and heterodyne experiments should be performed in the same experimental setup. Due to a significant loss in the laser output by going through the beam combiner of our collinear setup, we have opted to combine the non-collinear homodyne data with the collinear heterodyne phase result. It is to be noted that although the comparison of the local fields justifies such an approach where both the magnitude and the phase are practically identical given the small AOI difference in the non-collinear geometry bewteen the two probing beams, we note that there is a slight deviation in the higher incident angle region as shown in Figure 5.13. The local field correction factor for the heat-treated sample is shown in Figure 5.14.

5.3.5   Angle-resolved homodyne and heterodyne $\chi^{(2)}$ spectra

We first present the non-collinear p-polarized angle-dependent homodyne SFG spectra of the room temperature sample where, at each angle of incidence, the spectrum is collected from 2800 cm^{-1} to 3100 cm^{-1} as shown in Figure 5.15. If we recall the expression of the SFG signal in the reflection geometry in Eq. 2.1, the spectra as a function of the AOI cannot be directly compared as the $\sec^2 \theta_{\text{sfg}}$ term needs to first be considered. Additionally, any potential artifacts associated with AOI that arise from the rotation motor as well as misalignment are taken care of by considering a separate ssp experiment where a piece of gold is used in the exact same sample position. Since $\chi^{(2)}_{\text{ssp}}$ only contains a single tensor element of $\chi^{(2)}_{yyz}$ in $C_{\infty v}$ group. By carefully considering the local field contribution, we expect the resulting $\chi^{(2)}_{yyz}$ to be independent of the incident angle. Thus, we have attributed any additional lineshape deviation to the misalignment of the two beams walking off as the sample is rotated. Figure 5.16 shows the experimental determination of the lineshape deviation due to the rotation motor of the sample stage. From the initial alignment at 40° of AOI, there is a small deviation towards the latter part of the measurement starting at

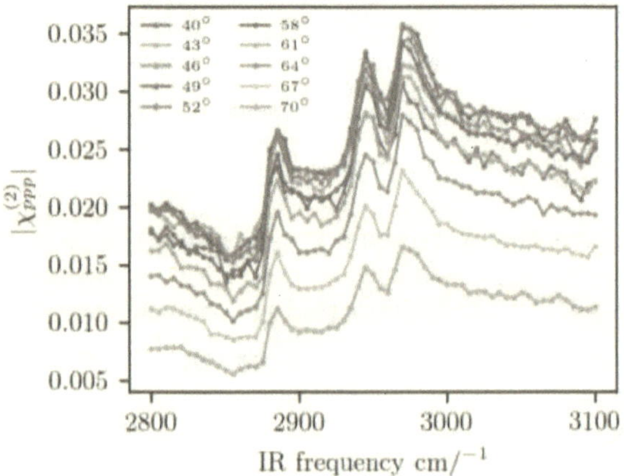

Figure 5.15: The magnitude of the ppp homodyne SFG spectra of BEP-Fe at room temperature as a function of angle of incidence. The spectra are taken from 2800 cm^{-1} to 3100 cm^{-1} with 100 laser shots averages at each frequency. The range of angle of incidence of the visible beam considered here is from 40° to 70°. All incident angle dependent terms relating to the reflective SFG signals are considered.

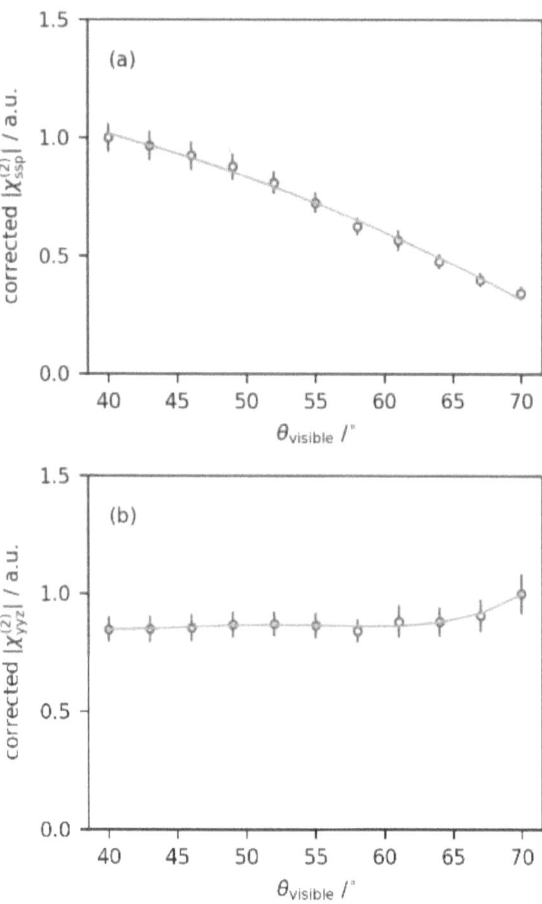

Figure 5.16: (a)The magnitude of the $\chi_{ssp}^{(2)}$ of air-gold interface as a function of angle of incidence at 2800 cm^{-1}; (b)The magnitude of the $\chi_{yyz}^{(2)}$ of air-gold interface as a function of angle of incidence at 2800 cm^{-1}.

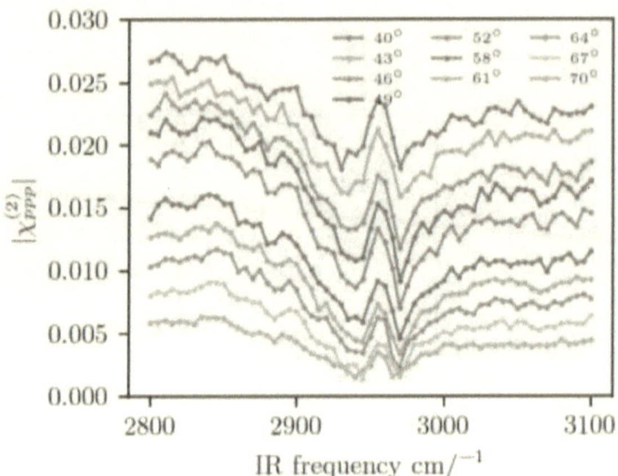

Figure 5.17: The magnitude of the ppp homodyne SFG spectra of 200 °C heat-treated BEP-Fe as a function of angle of incidence. The spectra are taken from 2800 cm^{-1} to 3100 cm^{-1} with 100 laser shots averages at each frequency. The range of angle of incidence of the visible beam considered here is from 40° to 70°. All incident angle dependent terms relating to the reflective SFG signals are consider.

roughly 60°. Such lineshape deviation is then considered in the BEP-iron spectra as the relative magnitude change from the AOI serves as an important constrain in the subsequent fitting.

By considering all contributions aside from the AOI dependent local field effect correction factors, the difference in the magnitude of $\chi^{(2)}_{\text{ppp}}$ in Figure 5.15 strictly comes from the relative difference in the local field contributions in various AOI. The AOI dependent $\chi^{(2)}_{\text{ppp}}$ of the 200 °C heat-treated sample is calculated in the same manner as shown in Figure 5.17. Although we can observe a significant spectral change from the two homodyne experiments alone, $\chi^{(2)}_{\text{ppp}}$ is a linear combination of the product of $\chi^{(2)}_{ijk}$ and its respective local field effect correction factor. Therefore, any subsequent interpretation of the spectral comparison should be done in such a way that the local field contributions do not come into play. As a result, we utilize the heterodyne measurement to extract the

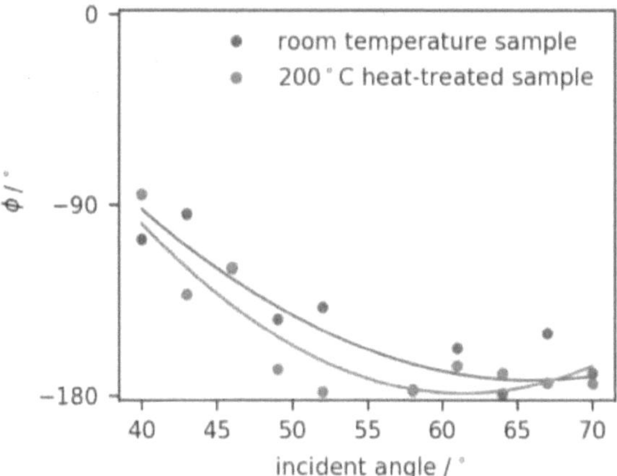

Figure 5.18: Heterodyne phase measurement at 2800 cm^{-1} of (blue) room temperature sample (red) 200 °C heat-treated sample. Experimental phase are fitted with second degree polynomials as the $\chi^{(2)}_{\mathrm{ppp,NR}}$ is a smooth function with respect to angle of incidence.

contributing tensor elements from the $\chi^{(2)}_{\mathrm{ppp}}$ in the same manner described in the previous chapter. Lastly, a brief note on the omission of the 55° spectra from both samples. The Brewser's angle of our phase reference (z-cut α quartz) of the heterodyne measurement is 55°. Consequently, no p-polarized beam can be reflected. Therefore, we are not able to perform our phase measurement at that AOI in a ppp polarization scheme.

The phase measurements are taken in the same AOI range as the homodyne measurements from 40° to 70° as shown in both Figure 5.18 and Figure 5.19 for 2800 cm^{-1} and 3100 cm^{-1}. A polynomial function is used to fit the phase as a function of the AOI given that the change in the non-resonant phase has to be a smooth function which can been seem by examining the phase of their respective local field shown in Figure 5.13 for the room temperature sample and Figure 5.14 for the 200 °C heat-treated sample.

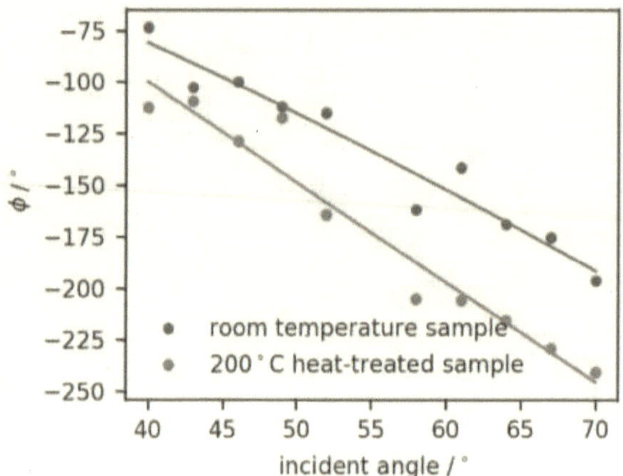

Figure 5.19: Heterodyne phase measurement at 3100 cm^{-1} of (blue) room temperature sample (red) 200 °C heat-treated sample. Experimental phase are fitted with second degree polynomials as the $\chi^{(2)}_{\text{ppp,NR}}$ is a smooth function with respect to angle of incidence.

5.3.6 SFG data fitting & analysis

5.3.6.1 $\chi^{(2)}_{\text{eff}}$ expression of the ppp measurement

We first present the linear combination of the achrial tensor elements for the effective $\chi^{(2)}_{\text{ppp}}$ as

$$
\begin{aligned}
\chi^{(2)}_{\text{ppp}} = {} & L_{zz}e_z L_{zz}e_z L_{zz}e_z \chi^{(2)}_{zzz} \\
& - L_{xx}e_x L_{xx}e_x L_{zz}e_z \chi^{(2)}_{xxz} \\
& + L_{zz}e_z L_{xx}e_x L_{xx}e_x \chi^{(2)}_{zxx} \\
& - L_{zz}e_z L_{xx}e_x L_{xx}e_x \chi^{(2)}_{xzx}.
\end{aligned}
$$

(5.12)

We note that from the local field factor plot, the zxx and xzx components are practically identical in both magnitude and phase for both room temperature sample and heat-treated sample. Thus, it is not possible for us to extract the two $\chi^{(2)}_{ijk}$ tensor elements within our measurement range of AOI. We can, therefore, describe the $\chi^{(2)}_{\text{ppp}}$ as

$$
\chi^{(2)}_{\text{ppp}} = L_{zz}e_z L_{zz}e_z L_{zz}e_z \chi^{(2)}_{zzz} - L_{xx}e_x L_{xx}e_x L_{zz}e_z \chi^{(2)}_{xxz} + L_{zz}e_z L_{xx}e_x L_{xx}e_x (\chi^{(2)}_{zxx} - \chi^{(2)}_{xzx}).
$$
(5.13)

Prior to fitting the experimental spectra, we first show an approximation of the magnitude and phase of the effective NR contribution in the case where there is significant resonant contribution relative to the non-resonance. We have modelled three cases where the NR contribution is either purely real, with a phase of $45°$, or purely imaginary shown in Figure 5.20. The model for all three spectra is fitted with three Lorentzian functions at 2875 cm^{-1}, 2940 cm^{-1}, and 2975 cm^{-1}. We note that with the presence of the resonant modes, the magnitude and the phase in the near-resonant region (at 2800 cm^{-1} and 3100 cm^{-1}) deviate from the true non-resonant values. Thus, the near-resonant experimental magnitude and phase of the $\chi^{(2)}_{\text{eff}}$ cannot be used to approximate the actual value. This is true even at the frequency of 3100 cm^{-1} where the closest nearby resonant mode is over 100 cm^{-1} away. The magnitude and phase of the $\chi^{(2)}_{\text{eff}}$ can be better approximated by taking the average from the pre- and post- resonance region regardless of the phase of the NR. We demonstrate such approach in Fig. 5.21 by using the $45°$ case from Fig. 5.20 as an example. In Fig. 5.21, we have demonstrated that by taking the averages of both magnitude and phase from the 2800 cm^{-1} and 3100 cm^{-1} values, we are able to approximate the true complex valued $\chi^{(2)}_{\text{NR}}$ as shown in Fig. 5.21 where the approximated $\chi^{(2)\text{NR}}$ value (in red dashed line) is close to that of the true $\chi^{(2)}_{\text{NR}}$ value (in blue dashed line).

By assuming the approximated non-resonant contribution of the interface, we can rewrite Eq. 5.13 by separating the resonant and non-resonant contribution as

$$
\begin{aligned}
\chi^{(2)}_{\text{PPP}} = |\chi^{(2)}_{\text{ppp,NR}}|e^{(i\phi_{\text{ppp,NR}})} &+ L_{zz}e_zL_{zz}e_zL_{zz}e_z\sum_q \frac{A_{zzz,q}}{\omega_q - \omega_{\text{IR}} - i\Gamma_q} \\
&- L_{xx}e_xL_{xx}e_xL_{zz}e_z\sum_q \frac{A_{xxz,q}}{\omega_q - \omega_{\text{IR}} - i\Gamma_q} + L_{zz}e_zL_{xx}e_xL_{xx}e_x\sum_q \frac{A_{zxx,q} - A_{xzx,q}}{\omega_q - \omega_{\text{IR}} - i\Gamma_q},
\end{aligned}
\tag{5.14}
$$

where the resonant contribution is represented as the summation of the Lorentzian functions to represent vibrational modes.

We can make further simplification of the expression of the $\chi^{(2)}_{\text{ppp}}$ by considering the relationship of the functional group orientation, molecular hyperpolarizability and the susceptibility tensor for molecular groups with C_{3v} and C_{2v} for the methyl and methlyene

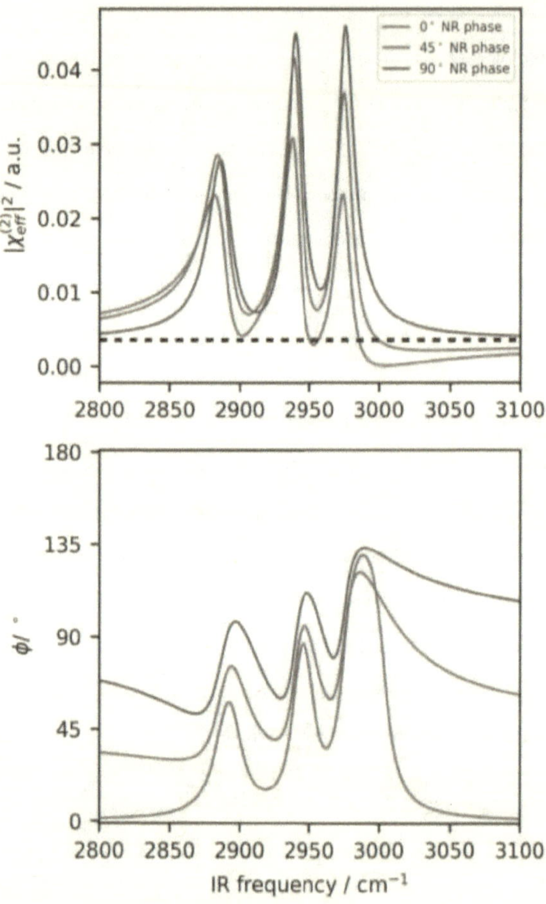

Figure 5.20: Values of the magnitude squared (top) and phase (bottom) spectra of the $\chi_{\mathrm{eff}}^{(2)}$ spectra for various non-resonant phase. The simulation illustrates the effect of resonant contributions to the overall magnitude and phase of the near-resonant region. The dashed line is the magnitude of the non-resonance.

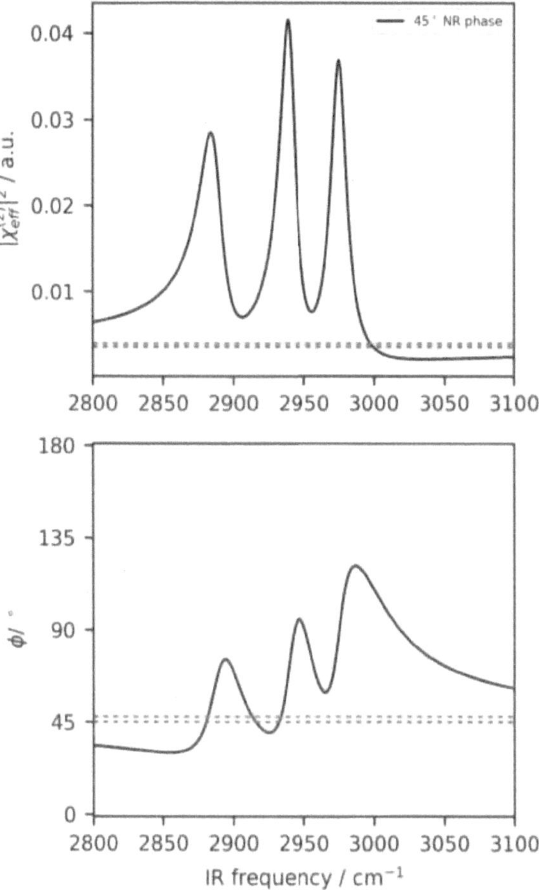

Figure 5.21: The magnitude squared (top) and phase (bottom) spectra of the $\chi^{(2)}_{\text{eff}}$ spectra for non-resonant phase of $45°$. The blue dashed line indicates the true $\chi^{(2)}_{\text{NR}}$ value whereas the red dashed line is the approximated $\chi^{(2)}_{\text{NR}}$ value.

vibrational modes, respectively. The definition of the molecular frame of (a,b,c) with respect to both the methyl and methlyene group is according to Ref. 54 where the c-axis is parallel to that of the carbon chain and the a-axis is in the same plane as one of the C–H bond while bisecting the other two hydrogen atoms for the methyl function group. For the methlyene group, the c-axis bisects the the two C–H bonds while being in parallel with the carbon chain. Furthermore, we define the rotation angle, ϕ to be the angle between x-axis of the lab frame and a-axis of the molecular frame when ab-plane equals to xy-plane. The twist angle, ψ is defined as the angle between the ab-plane in the molecular frame and the xy-plane in the lab frame. Finally, the tilt angle, θ is defined as the angle between the z-axis of the lab frame to the c-axis of the molecular frame. We then can define the Euler transformation matrix $\mathbf{D}$ between the laboratory frame of (x,y,z) to the molecular frame of (a,b,c) as[216,217]

$$\mathbf{D}(\theta,\phi,\psi)$$
$$= \begin{bmatrix} \cos\psi\cos\phi - \cos\phi\sin\phi\sin\psi & -\sin\psi\cos\phi - \cos\theta\sin\phi\cos\psi & \sin\theta\sin\phi \\ \cos\psi\sin\phi + \cos\phi\cos\phi\sin\psi & -\sin\psi\sin\phi + \cos\theta\cos\phi\cos\psi & -\sin\theta\cos phi \\ \sin\theta\sin\psi & \sin\theta\cos\psi & \cos\theta \end{bmatrix}.$$
$$(5.15)$$

First, we consider the C_{3v} symmetry group, by assuming a rotationally isotropic distribution about the c-axis where the $A_{ijk}(\theta)$ is obtained through the integration over both rotational and twist angles. Thus, the four unique non-zero $\chi_{ijk}^{(2)}$ can be expressed as

$$\chi_{xxz,ss}^{(2)} = \frac{1}{2}N\alpha_{ccc}^{(2)}[(1+R)\langle\cos\theta\rangle - (1-R)\langle\cos^3\theta\rangle]$$

$$\chi_{xzx,ss}^{(2)} = \chi_{zxx,ss}^{(2)} = \frac{1}{2}N\alpha_{ccc}^{(2)}(1-R)[\langle\cos\theta\rangle - \langle\cos^3\theta\rangle]$$

$$\chi_{zzz,ss}^{(2)} = N\alpha_{ccc}^{(2)}[R\langle\cos\theta\rangle + (1-R)\langle\cos^3\theta\rangle]$$

$$\text{and} \qquad\qquad (5.16)$$

$$\chi_{xxz,as}^{(2)} = -N\alpha_{aca}^{(2)}(\langle\cos\theta\rangle - \langle\cos^3\theta\rangle)$$

$$\chi_{xzx,as}^{(2)} = \chi_{zxx,as}^{(2)} = N\alpha_{aca}^{(2)}(\langle\cos^3\theta\rangle)$$

$$\chi_{zzz,as}^{(2)} = 2N\alpha_{aca}^{(2)}(\langle\cos\theta\rangle - \langle\cos^3\theta\rangle),$$

where N is the surface number density, $\alpha_{lmn}^{(2)}$ is the molecular hyperpolarizability, R is the ratio between $\alpha_{aac}^{(2)}$ amd $\alpha_{ccc}^{(2)}$ and the subscript of *as* refers to the antisymmetric mode, and the subscript of *ss* refers to the symmetric mode.

Similarly, By assuming the free rotation of the H-C-H plane about the c-axis where there is an isotropic distribution of the twist angle, the C_{2v} symmetry for the methlyene functional group can be described in a similar manner as

$$\chi_{xxz,ss}^{(2)} = \frac{1}{4}N(\alpha_{aac}^{(2)} + \alpha_{bbc}^{(2)} + 2\alpha_{ccc}^{(2)})\langle\cos\theta\rangle$$
$$+ \frac{1}{4}N(\alpha_{aac}^{(2)} + \alpha_{bbc}^{(2)} - 2\alpha_{ccc}^{(2)})\langle\cos^3\theta\rangle$$
$$\chi_{xzx,ss}^{(2)} = \chi_{zxx,ss}^{(2)} = -\frac{1}{4}N(\alpha_{aac}^{(2)} + \alpha_{bbc}^{(2)} - 2\alpha_{ccc}^{(2)})(\langle\cos\theta\rangle - \langle\cos^3\theta\rangle)$$
$$\chi_{zzz,ss}^{(2)} = \frac{1}{2}N(\alpha_{aac}^{(2)} + \alpha_{bbc}^{(2)})\langle\cos\theta\rangle - \frac{1}{2}N(\alpha_{aac}^{(2)} + \alpha_{bbc}^{(2)} - 2\alpha_{ccc}^{(2)})\langle\cos^3\theta\rangle \quad (5.17)$$

and

$$\chi_{xxz,as}^{(2)} = -\frac{1}{2}N\alpha_{aca}^{(2)}(\langle\cos\theta\rangle - \langle\cos^3\theta\rangle)$$
$$\chi_{xzx,as}^{(2)} = \chi_{zxx,as}^{(2)} = \frac{1}{2}N\alpha_{aca}^{(2)}\langle\cos^3\theta\rangle$$
$$\chi_{zzz,as}^{(2)} = N\alpha_{aca}^{(2)}(\langle\cos\theta\rangle - \langle\cos^3\theta\rangle).$$

By considering both Eq. 5.16 and 5.17, we note that xxz and xzx are equivalent for both functional group stretching modes. Thus, Eq. 5.14 becomes

$$\chi_{PPP}^{(2)} = |\chi_{PPP,NR}^{(2)}|e^{(i\phi_{PPP,NR})} + L_{zz}e_zL_{zz}e_zL_{zz}e_z\sum_q\frac{A_{zzz,q}}{\omega_q - \omega_{IR} - i\Gamma_q}$$
$$- L_{xx}e_xL_{xx}e_xL_{zz}e_z\sum_q\frac{A_{xxz,q}}{\omega_q - \omega_{IR} - i\Gamma_q}, \quad (5.18)$$

with two tensor elements contributing to the overall $\chi_{PPP}^{(2)}$ resonant response.

5.3.6.2 Phase relationships between vibrational modes

Prior to fitting our experimental spectra, the absolute sign constraint of the amplitude of each mode needs to be taken into consideration. Here we discuss the absolute sign constraint between the symmetric and antisymmetric stretching mode of the methyl function group. For simplicity, we use the methyl hyperpolarizability ratio, $R = 3.4$ for ethanol and longer chain primary alcohols.[52,54] In addition, similar to the situation

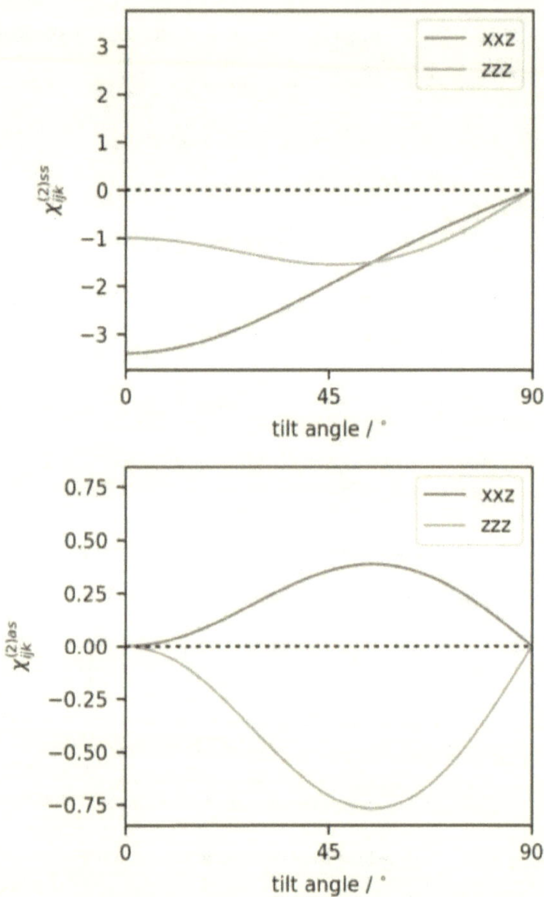

Figure 5.22: Calculation of the amplitude of the $\chi^{(2)}_{xxz}$ and $\chi^{(2)}_{zzz}$ terms for the methyl mode as a function of the tilt angle, θ.

discussed in Ref. 52 where the sign of the $\alpha_{ccc}^{(2)}$ and $\alpha_{aca}^{(2)}$ are negative for the methyl group to be pointing towards the environment side while using the same Lorentzian function convention as shown in Eq. 2.7, we can deduce the absolute sign of the $\chi_{xxz}^{(2)}$ and $\chi_{zzz}^{(2)}$ as shown in Figure 5.22. Note that the magnitude of the $\alpha_{ccc}^{(2)}$ and $\alpha_{aca}^{(2)}$ are not considered here. Therefore, only their relative sign should be compared with each $\chi_{ijk}^{(2)}$ tensor element. No conclusion should be drawn comparing the magnitude between the symmetric and antisymmetric modes. Finally, from the above analysis we can conclude $A_{zzz,ss} < 0$, $A_{zzz,as} < 0$, $A_{xxz,ss} < 0$ and $A_{xxz,as} > 0$.

A similar argument can be made for the methylene group where we use

$$R = \frac{\alpha_{aac}^{(2)} + \alpha_{bbc}^{(2)}}{2\alpha_{ccc}^{(2)}} = 1.0$$

from the bond additive model when considering the CH_2 bond angle of 109.5°. An R value of unity implies that $\alpha_{aac}^{(2)} + \alpha_{bbc}^{(2)} - 2\alpha_{ccc}^{(2)} = 0$ and is generally true for the methylene group.[54] Thus, we can rewrite Eq. 5.17 while considering only the tensor elements of xxz and zzz as

$$\begin{aligned}
\chi_{xxz,ss}^{(2)} &= \frac{1}{4}N(\alpha_{aac}^{(2)} + \alpha_{bbc}^{(2)} + 2\alpha_{ccc}^{(2)})\langle\cos\theta\rangle \\
\chi_{zzz,ss}^{(2)} &= \frac{1}{2}N(\alpha_{aac}^{(2)} + \alpha_{bbc}^{(2)})\langle\cos\theta\rangle \\
\chi_{xxz,as}^{(2)} &= -\frac{1}{2}N\alpha_{aca}^{(2)}(\langle\cos\theta\rangle - \langle\cos^3\theta\rangle) \\
\chi_{zzz,as}^{(2)} &= N\alpha_{aca}^{(2)}(\langle\cos\theta\rangle - \langle\cos^3\theta\rangle).
\end{aligned} \quad (5.19)$$

By considering the bond additive model along with $R = 1$ for the methlyene group, we can calculate the four contributing hyperpolarizabilities as

$$\begin{aligned}
\alpha_{aac}^{(2)} &= \frac{G_a\alpha_{CH}^{(2)}}{\omega_a}[(1+r) - (1-r)\cos\tau]\cos\frac{\tau}{2} \\
\alpha_{bbc}^{(2)} &= \frac{2G_a\alpha_{CH}^{(2)}}{\omega_a}r\cos\frac{\tau}{2} \\
\alpha_{ccc}^{(2)} &= \frac{G_a\alpha_{CH}^{(2)}}{\omega_a}[(1+r) + (1-r)\cos\tau]\cos\frac{\tau}{2} \\
\alpha_{aca}^{(2)} &= \frac{G_b\alpha_{CH}^{(2)}}{\omega_a}[(1-r)\sin\tau]\sin\frac{\tau}{2},
\end{aligned} \quad (5.20)$$

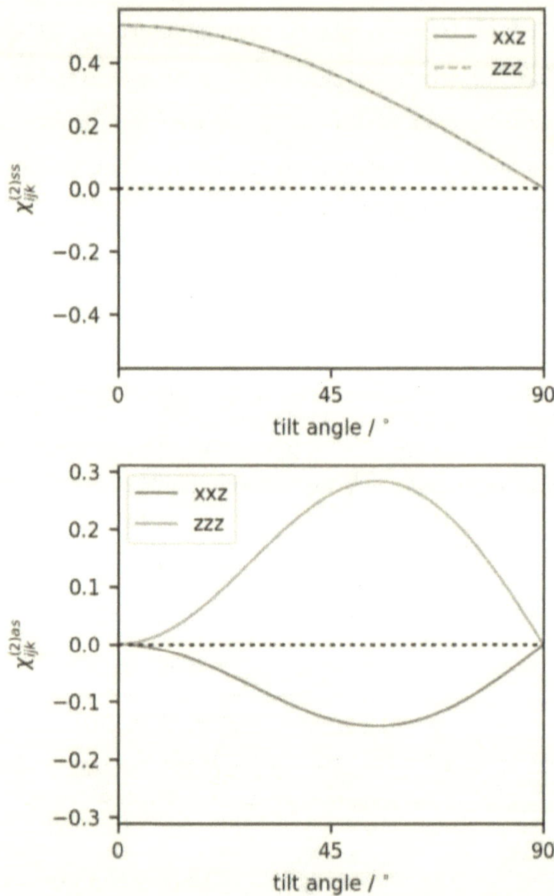

Figure 5.23: Theoretical calculation of the amplitude of the $\chi^{(2)}_{xxz}$ and $\chi^{(2)}_{zzz}$ terms for the methylene mode as a function of the tilt angle, θ.

where τ is the CH_2 bond angle, G_a and G_b are the inverse effective mass for the symmetric and antisymmetric normal modes at frequency, ω, $\alpha_{CH}^{(2)}$ is the reference hyperpolarizability for a single C–H bond, and r is the single C–H bond polarizability derivative ratio which can be obtained from the Raman depolarization ratio.[54] By using $r = 0.14$ as in Refs. 218 and 219, and setting $\alpha_{CH}^{(2)}$ as unity, we have calculated the relative sign between the two modes for both xxz and zzz as shown in Figure 5.23. In the case of the methylene calculation, only the relative sign should be considered as the calculation does not consider the absolute sign of the $\alpha_{CH}^{(2)}$ nor the frequency dependence of each mode. Thus, we can only conclude that the sign of the vibrational amplitudes for the methylene group can be either ($A_{zzz,ss} > 0$, $A_{zzz,as} > 0, A_{xxz,ss} > 0, A_{xxz,as} < 0$) or ($A_{zzz,ss} < 0, A_{zzz,as} < 0, A_{xxz,ss} < 0, A_{xxz,as} > 0$). Note that these relations hold independently of the measured phase. When experimental phase data is available, only one of these sets will be true. That distinction is to be resolved in the spectral fitting when the phase of NR is known.

5.3.6.3 Data fitting of the room temperature sample

Now that the non-resonant phase and relationship between the vibrational mode amplitudes has been addressed, we present the fitting of the combination of the homodyne and heterodyne SFG measurements for the BEP-Fe interface in room temperature as shown in Figure 5.24 and its phase shown in Figure 5.25. Each amplitude of the Lorentzian function is obtained by a global fitting across all 10 angles along with the respective complex valued $\chi_{ppp,NR}^{(2)}$ obtained from the heterodyne phase measurement. We have assume a simple model for the vibrational modes by considering that all the methyl groups on the BEP are in the same environment while the H-C-H plane of the methylene group freely rotates about its c-axis. As a result, and in consideration of the appearance of the intensity data, we have considered 5 vibrational modes, each contributing A_{xxz} and A_{zzz} to the Lorentzian lineshape. These modes are identified in Table 5.3.

The fitting is done by first optimizing the frequency and the line width of each mode,

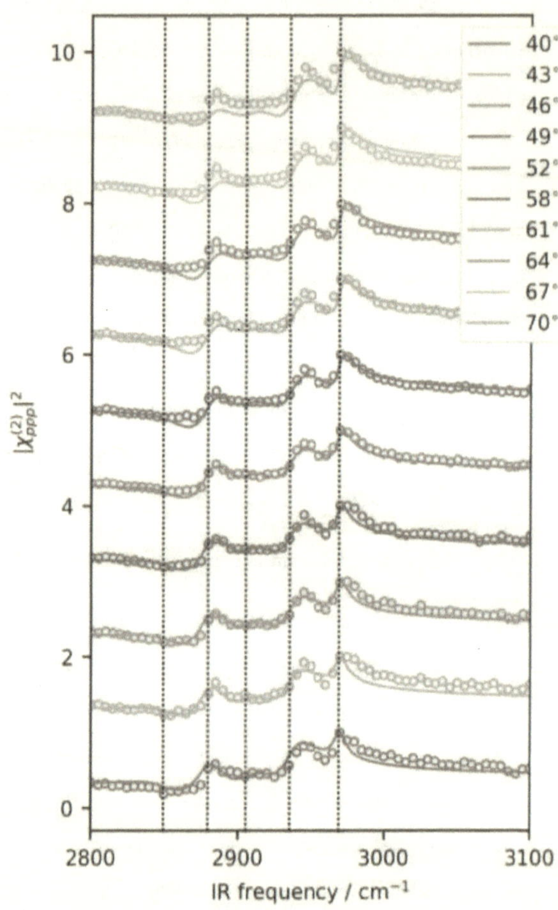

Figure 5.24: Fitting result of the multi-AOI homodyne and heterodyne SFG measurements of the BEP-Fe interface of the room temperature sample, plotting $|\chi^{(2)}_{\mathrm{ppp}}|^2$ with the experimental values in dots and the fitting result in lines. The AOI-dependent $|\chi^{(2)}_{\mathrm{ppp}}|^2$ spectra shown here are normalized and scaled in order to separate between each incident angle for illustration purposes. The relative magnitude between each spectrum as a function of the incident angle is in fact, an important constrain for the fit.

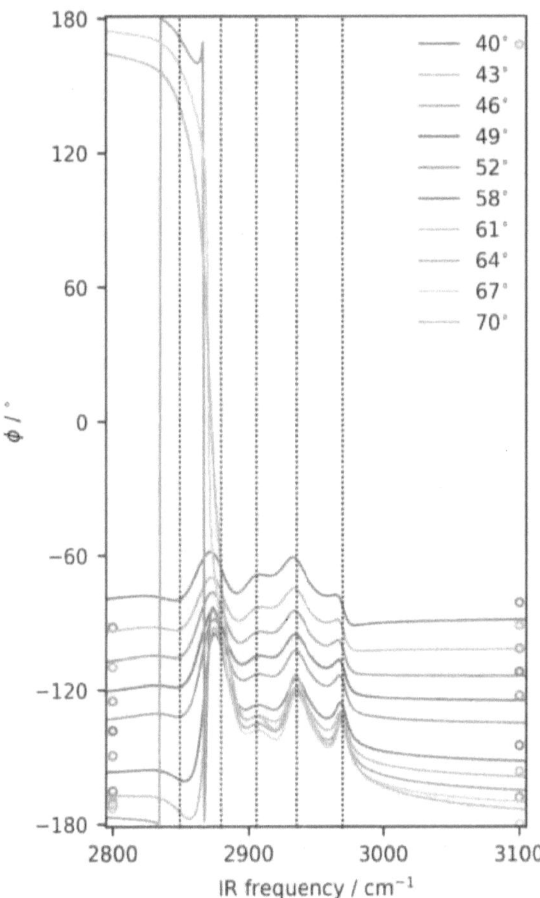

Figure 5.25: Resulting phase of the $\chi_{ppp}^{(2)}$ as a function of the incident angle from the fitting of the room temperature sample in solid lines. The experimental measured phase of the $\chi_{ppp}^{(2)}$ as a function of the incident angle from the fitting for the room temperature sample is plotted with points at 2800 cm^{-1} and 3100 cm^{-1}.

ω_q / cm^{-1}	Assignment	Γ_q	$A_{xxz,q}$	$A_{zzz,q}$
2850	CH$_2^{ss}$	19.8	0.542	0.707
2879	CH$_3^{ss}$	13.1	-1.901	-0.668
2905	CH$_2^{as}$	17.9	-1.46	1.14
2936	CH$_3^{FR}$	12.8	-1.46	-1.09
2970	CH$_3^{as}$	5.3	-0.30	0.05

Table 5.3: Tabulation of the resulting fitting parameters from the global fitting of the combination of multiple AOI homodyne and heterodyne SFG spectra for the room temperature sample.

then by using a gradient decedent optimizier (truncated Newton algorithm), we obtained the amplitude of each Lorentzian function. The fitting result is shown in Table 5.3. The resulting $\chi_{ijk}^{(2)}$ spectra are shown in Figure 5.26.

5.3.6.4 Room temperature BEP structural analysis

By taking the ratio between $A_{xxz,q}$ and $A_{zzz,q}$ following Eq. 5.16, we can calculate the tilt angle, θ, by comparing to the experimentally determined ratio of $A_{xxz,q}/A_{zzz,q}$ described as

$$f(\theta) = \frac{\chi_{xxz,ss}^{(2)}}{\chi_{zzz,ss}^{(2)}} = \frac{\frac{1}{2}[(1+R)\langle\cos\theta\rangle - (1-R)\langle\cos^3\theta\rangle]}{R\langle\cos\theta\rangle + (1-R)\langle\cos^3\theta\rangle} \tag{5.21}$$

which is shown in Figure 5.27 for a narrow distribution of tilt angles with an experimental ratio of $f = 2.85$. Therefore, we have determined that the preliminary average methyl group tilt angle of the BEP structured on iron at room temperature is ca. 15°. In other words, the methyl group is likely to structure 15° away from the surface normal on average with an isotropic distribution about the xy-plane in the laboratory frame. This conclusion also assumes that the H-C-H plane of the methylene freely rotates about its c-axis, which is generally a valid assumption when no other function group or atom is attached to the carbon chain.[54]

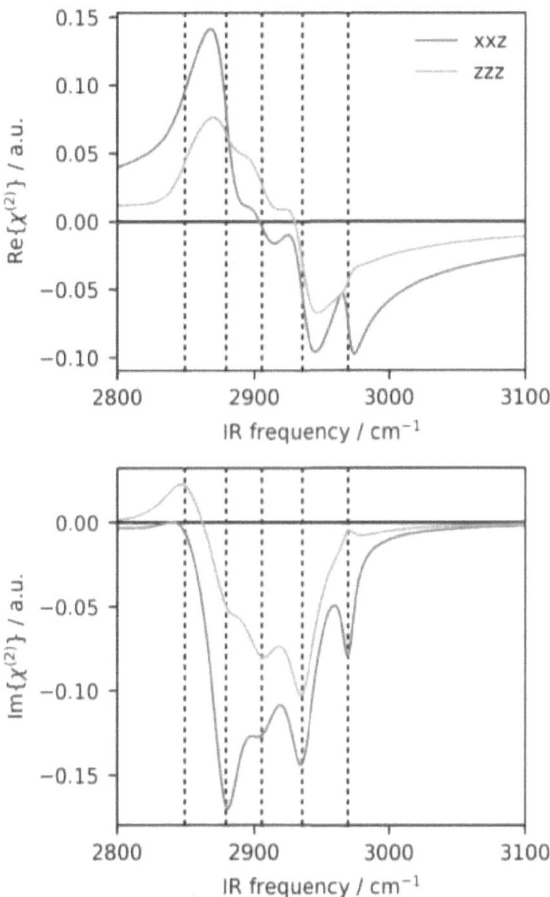

Figure 5.26: Complex $\chi^{(2)}$ spectra of the *xxz* and *zzz* tensor elements for the BEP-Fe interface at room temperature.

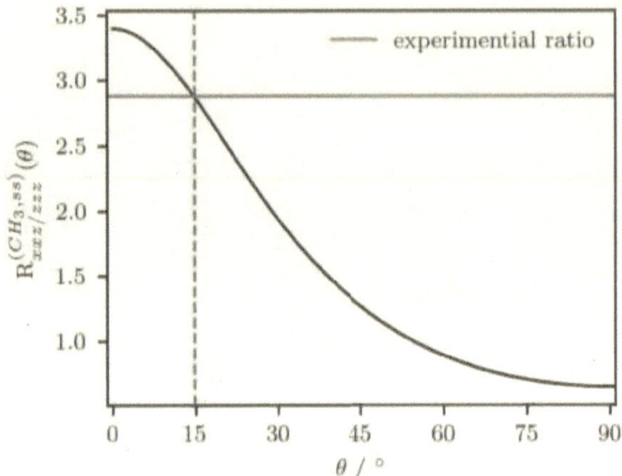

Figure 5.27: Predicted amplitude ratio of $A_{xxz,q}/A_{zzz,q}$ as a function of the tilt angle in black with a R value of 3.4, assuming a narrow distribution of tilt angles. The experimentally determined amplitude ratio is plotted in red solid line and the corresponding tilt angle is indicated by the red dashed line.

5.3.6.5 Data fitting of the heat-treated sample

We have applied the same resonance frequency and line width parameters from the room temperature sample fitting result to the 200 °C heat-treated sample spectra as shown in Figure 5.28 for the intensity and its resulting phase shown in Figure 5.29. We are unable to account for the spectral features of the spectra for the heat-treated sample when the same constraints are applied. There are a few possible reasons to explain the deviation between the fitting and the experimental homodyne spectra. First, it is possible that the sign constrains described in the above section are no longer valid given the H-C-H plane of the methlyene group can no longer freely rotate about its c-axis after the re-orientation. This implies the twist angle cannot be integrated to afford Eq. 5.17. It is therefore possible that no frequency shift or peak broadening has occurred, but the sign constraints applied to the fitting routine needing to be relaxed. Further investigation of this approach will need to be conducted to examine this hypothesis. On the other hand,

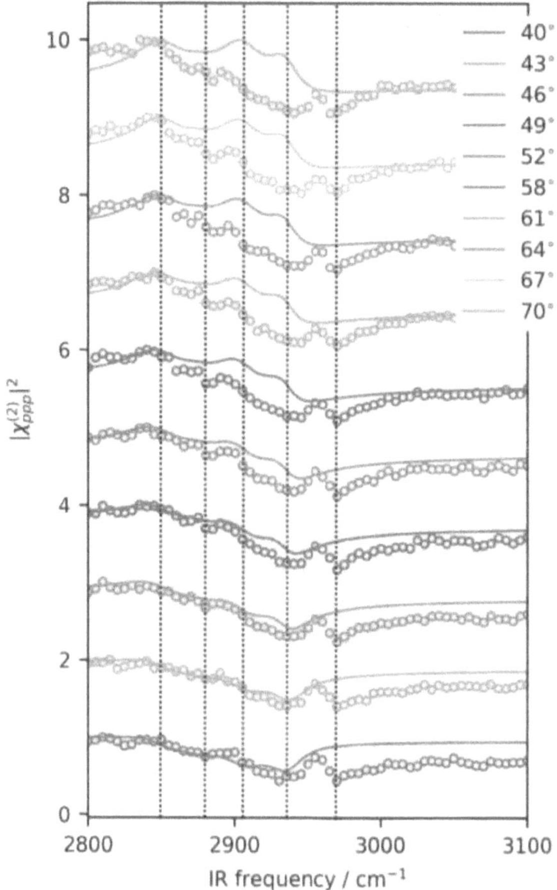

Figure 5.28: Fitting result of the multi-AOI homodyne and heterodyne SFG measurements of the BEP-Fe interface of the heat-treated sample by using the same fitting constraints as the room temperature sample data, plotting $|\chi^{(2)}_{\mathrm{ppp}}|^2$ with the experimental values in dots and the fitting result in lines.

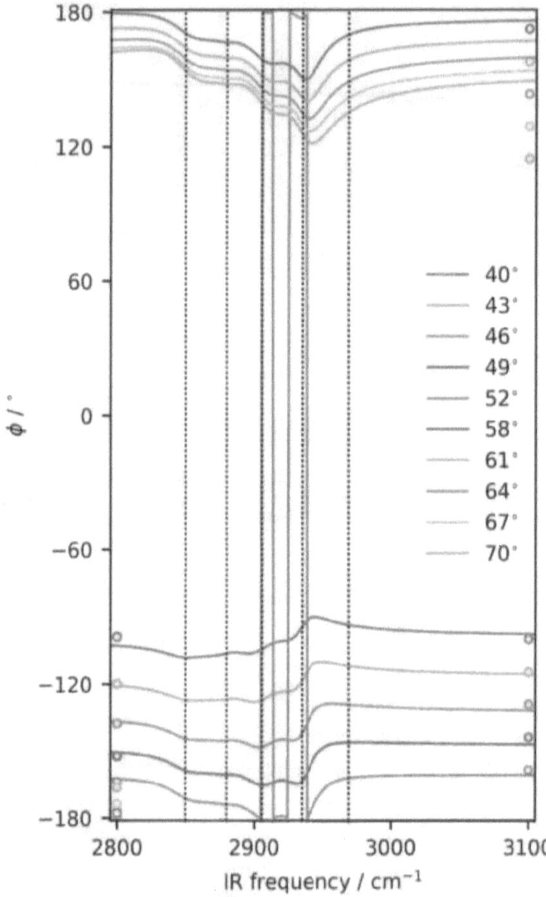

Figure 5.29: Resulting phase of the $\chi^{(2)}_{\mathrm{ppp}}$ as a function of the incident angle from the fitting of the heat-treated sample in solid lines. The experimental measured phase of the $\chi^{(2)}_{\mathrm{ppp}}$ as a function of the incident angle from the fitting for the heat-treated sample in points.

it is also possible that this simple model does not account for the potential frequency shift or peak broadening due to the re-orientation of the molecules. In cases where the peak broadening is expected, the inhomogeneous component, Γ_G from Eq. 2.10 may need to be considered when substituting the Lorentzian function to the Faddeeva function as mentioned in Chapter 2. Future adoption of a theoretical calculation of the normal mode may aid in assigning the vibrational features of the spectra.

Given the similarity of the phase of the $\chi^{(2)}_{\mathrm{ppp}}$ in the NR region as well as resemblance of the lineshape of their respective local field correction factor, we can conclude that the spectral changes observed in the homodyne spectra are likely due to the structural changes of the BEP. We present one additional fitting attempt for the heat-treated sample where we introduce a sign flip for the amplitude of the methyl modes as shown in Fig. 5.30 with the resulting phase plotted in Fig. 5.31. Switching the sign of the methyl *ss* and *as* modes results in an improved fitting output at the higher frequency region where we expect the methyl anti-symmetric stretch to be located. However, given the range of the approximated NR phase is close to $180°$ as a function of the angle of incidence, we have observed a sign change of the $\chi^{(2)}_{\mathrm{ppp,eff}}$ in the intensity spectra starting from the $52°$ incident angle spectrum which does not agree with our experimental homodyne spectra.

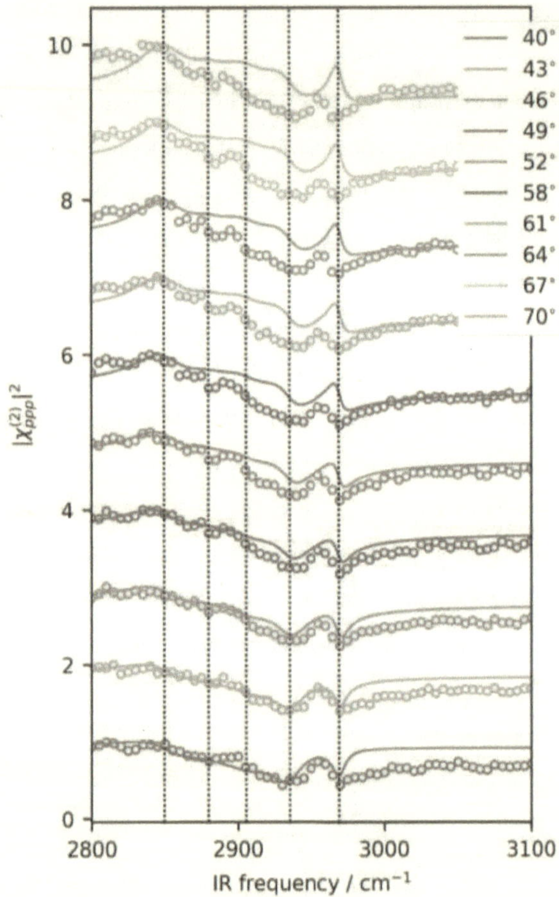

Figure 5.30: Fitting result of the multi-AOI homodyne and heterodyne SFG measurements of the BEP-Fe interface of the heat-treated sample by using the same fitting constraints but opposite sign for the methyl modes as the room temperature sample data, plotting $|\chi^{(2)}_{\mathrm{ppp}}|^2$ with the experimental values in dots and the fitting result in lines.

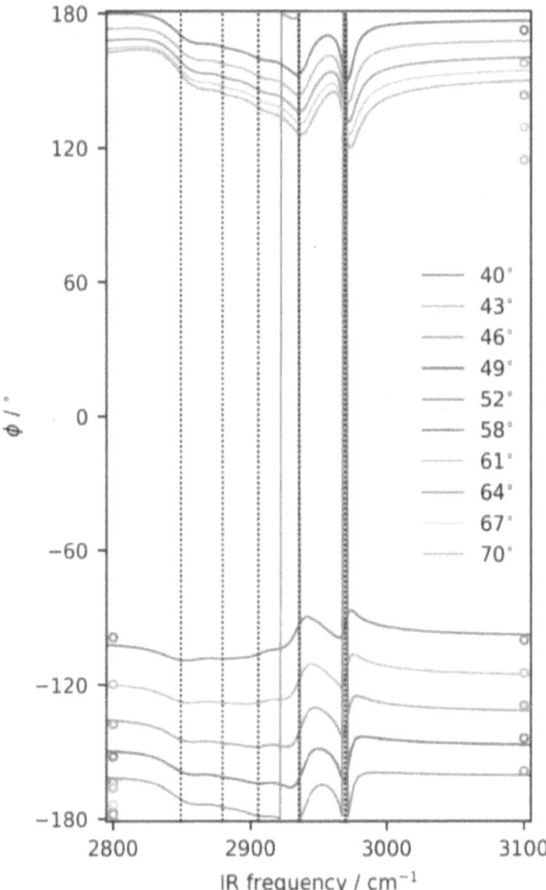

Figure 5.31: Resulting phase of the $\chi^{(2)}_{\mathrm{ppp}}$ as a function of the incident angle from the fitting of the heat-treated sample with an opposite sign constraint for the methyl mode in solid lines. The experimental measured phase of the $\chi^{(2)}_{\mathrm{ppp}}$ as a function of the incident angle from the fitting for the heat-treated sample in points.

5.4   Conclusion

We have applied our approach of the decomposition of the $\chi^{(2)}$ tensor elements by using the incident angle-resolved heterodyne SFG measurement to afford the complex $\chi^{(2)}$ for all contributing tensor elements in the vibrational resonant region. This approach is especially useful when only one experimental beam polarization scheme is accessible due to the surface selection rule of the metal substrate. In addition, this method does not suffer from any potential polarization bias from the optic given that it is only performed in one polarization combination. In addition, we have reported the experimentally determined structural orientation of an alkyl phosphate surfactant on iron surface with an average tilt angle of $15°$ for the methyl function group at room temperature. Additional theoretical study such as a normal mode analysis of the molecule may be required to confirm our experimental result. Further investigation of the elevated temperature sample result will be needed in order to compare the potential structural changes of the molecule adsorbed on the iron surface. In addition, we have also obtained the preliminary corrosion inhibition efficiency change of the alkyl phosphate at elevated temperature.

Chapter 6

Conclusions

6.1 Summary

I first illustrated how phase-sensitive sum-frequency generation spectroscopy may be used to determine the polar orientation of organic species on metals, in cases where there is a significant electronic contribution to the second-order signal. It turns out that traditional direct heterodyne schemes—as would be applied to the same molecules on dielectric substrates—are challenging to use here as a result of the large resonant/non-resonant amplitude ratio that diminishes the phase contrast observed in tuning through the vibrational modes. This is demonstrated in a variety of experimental surfaces that illustrate all limiting cases. I proposed a scheme that can overcome this challenge, and thereby determine the chemical functional group orientation through a combination of the homodyne spectrum, and at least some phase information from a heterodyne approach.

I further developed a technique by which heterodyne-detected sum-frequency generation spectroscopy is performed at multiple angles of incidence in order to decompose components of the second-order susceptibility tensor when all beams are polarized parallel to the plane of incidence. As an illustration, I studied the non-vibrationally resonant gold response. I benchmarked the results by comparing with measurements obtained in a polarization scheme that isolates a single element of the susceptibility tensor. This technique is particularly valuable in the case of metal substrates, where the surface selection rule often prevents spectra from being acquired in multiple beam polarizations.

Lastly, I applied the multiple incident angle measurement with the combination of the homodyne and heterodyne SFG to study a corrosion-inhibiting surfactant on an iron surface. To enable this study, I have also constructed a rotating compensator ellipsometer in the mid-infrared to obtain accurate complex refractive index data of the substrate for the calculation of the local field correction factors. The determination of the structural orientation of the surfactant on iron is carried out by fitting the experimentally determined complex-valued $\chi_{\mathrm{NR}}^{(2)}$ and the homodyne spectra as a function of the incident angle. The fitting affords the separation of contributing complex-valued $\chi^{(2)}$ elements as a function of frequency. These elements are then used to common on the structure of the surfactant on the iron surface.

6.2 Perspective and Future Work

The work outlined in this book focuses on the SFG technique and method development for organic–metal interfacial studies. A better understanding of the $\chi_{\mathrm{NR}}^{(2)}$ contribution from the metal substrate along with its effect on the overall $|\chi_{\mathrm{eff}}^{(2)}|^2$ line shape is crucial to enable further quantitative analysis of the spectra. The primary goal of this study was the examination of the $\chi_{\mathrm{eff}}^{(2)}$ decomposition of the p-polarized measurement in multiple angles of incidence. Although, I have presented a method to obtain the complex $\chi_{ijk}^{(2)}$ from decomposition of the $\chi_{\mathrm{ppp,eff}}^{(2)}$, the experimental phase of the $\chi_{\mathrm{ppp,eff}}^{(2)}$ in the vibrational resonant region is much more desirable as it affords the imaginary $\chi^{(2)}$ spectra of the $\chi_{ijk}^{(2)}$ without any assumption on the vibrational modes and such technique does not require any spectral fitting, only the decomposition of the tensor elements. Future improvement in terms of the acquisition time for the multi-angle of incidence measurement is necessary for subsequent studies to perform such technique as a function of frequency. An adoption of the technique to the broadband system may be of interest as far as the acquisition time is concerned.

In addition, the treatment of multiple incident angle spectra relies heavily on the calculated local field correction factor. Thus, the determination of optical constants is crucial in obtaining the true $\chi^{(2)}$ spectra. This is especially true for substrates such as iron that are highly reactive. I have observed a significant variation between published refractive index results that also appear to be highly dependent on the species of iron, and the oxide composition. It is for that reason that I have spent a significant amount of time to construct my own mid-IR ellipsometer to determine these optical constants, and have performed elemental analysis from the EDX to determine the oxide composition. Such approach ensures the accuracy of our orientation analysis from the SFG data, and I feel that for substrate such as iron, it is indispensable. In addition, measurements such as Mössbauer spectroscopy may be of interest in the future to determine the exact oxide composition for a better understanding of the substrate which in turns can validate our chose of substrate model when calculating the complex refractive index from the ellipsometry data.

For the surfactant–iron interface, the phase of the $\chi^{(2)}_{\mathrm{ppp,eff}}$ in the vibrational resonant region as a function of incident angle is a desirable piece of confirmation in order to further validate the current spectral fitting result. A important next step is to move from the exposed interface (air–metal) to buried interfaces (oil–metal or water–metal), as they can better represent real-life conditions of how these surfactants are being deployed as corrosion inhibitors. Recently, a proof of concept study of the heterodyne measurement of a buried interface has been published[52] for the water-metal interface. Future studies of the project should focus on the buried interface with a variety of environmental conditions to better match that of the real-life conditions of these surfactants. Parameters such as concentration of the surfactant, flow rate of the liquid phase, and temperature are of significant interest.

References

[1] Fontana, M. G.; Green, N. *Corrosion Engineering;* McGraw-Hill: New York, USA, 3rd ed.; 1978.

[2] McCafferty, E. *Introduction to Corrosion Science;* Springer: New York, USA, 2019.

[3] Ahmad, Z. *Principles of Corrosion Engineering and Corrosion Control;* Butterworth-Heinemann: Oxford, UK, 1st ed.; 2006.

[4] Shipilov, S., Ed.; *What Corrosion Costs Canada: Or, Can We Afford to Ignore Corrosion?,* ; The Royal Society of Chemistry: Cambridge, UK, 1999.

[5] Uhling, H. *Corrosion* **1952,** *6,* 29–33.

[6] Derra, S. *R D Manag.* **1970,** None, 29.

[7] Eckerman, I. *THE BHOPAL SAGA Causes and Consequences of the World's Largest Industrial Disaster;* University Press (India) Private Limited: Telangana, India, 2005.

[8] Arkema, "Arkema Technical Note QC", Technical Report 173, Arkema S.A., 2001.

[9] Malik, M. A.; Ali, M.; Nabi, F.; Al-Thabaiti, S. A.; Khan, Z. *Int. J. Electrochem. Sci.* **2011,** *6,* 1927–1948.

[10] Zhu, Y.; Free, M.; Woollam, R.; Durnie, W. *Prog. Mater. Sci.* **2017,** *90,* 159–223.

[11] Casford, M. T. L.; P. B. Davies, T. D. S.; Bracchi, G. L. *Tribol. Lett.* **2016,** *62,* 11.

[12] Campana, M.; Teichert, A.; Clarke, S.; Steitz, R.; Webster, J. R. P.; Zarbakhsh, A. *Langmuir* **2011,** *27,* 6085–6090.

[13] Wood, M. H.; Casford, M. T.; Steitz, R.; Zarbakhsh, A.; Welbourn, R. J. L.; Clarke, S. M. *Langmuir* **2016,** *32,* 534–540.

[14] Coletti, G.; Hewitt, G. F. *Crude Oil Fouling: Deposit Characterization, Measurements, and Modeling;* sean: sean, 2014.

[15] Boyd, R. W. *Nonlinear Optics;* Academic Press: Boston, 1992.

[16] Shen, Y. R. *The Principles of Nonlinear Optics;* John Wiley & Sons: New York, 1984.

[17] Franken, P. A.; Hill, A. E.; Peters, C. W.; Weinreich, G. *Phys. Rev. Lett.* **1961,** *7,* 118–120.

[18] Bain, C. D. *J. Chem. Soc. Faraday Trans.* **1995,** *91,* 1281–1296.

[19] Rocha-Mendoza, I.; Yankelevich, D. R.; Wang, M.; Reiser, K. M.; Frank, C. W.; ; Knoesen, A. *Biophys. J.* **2007,** *93,* 4433–4444.

[20] Liljeblad, J. F. D.; Tyrode, E. *J. Phys. Chem. C* **2012,** *116,* 22893–22903.

[21] Zhang, C. *Appl. Spectrosc.* **2017,** *71,* 1717–1749.

[22] Takeshita, N.; Okuno, M.; aki Ishibashi, T. *J. Phys. Chem. C* **2017,** *121,* 25206–25214.

[23] Ge, H.-H.; Hochstrasser, R. M. *Phys. Chem. Comm.* **2002,** *5,* 17–26.

[24] Hoffmann, H.; Mayer, U.; Brunner, H.; Krischanitz., A. *Vib. Spectrosc.* **1995,** *8,* 151–157.

[25] Matsuzaki, K.; Nihonyanagi, S.; Yamaguchi, S.; Nagata, T.; Tahara, T. *J. Phys. Chem. Lett.* **2013**, *4*, 1654–1658.

[26] Zhang, C.; Jasensky, J.; Chen, Z. *Anal. Chem.* **2015**, *87*, 8157–8164.

[27] Hall, S. A.; Jena, K. C.; Covert, P. A.; Roy, S.; Trudeau, T. G.; Hore, D. K. *J. Phys. Chem. B* **2014**, *118*, 5617–5636.

[28] Rao, Y.; Comstock, M.; Eisenthal, K. B. *J. Phys. Chem. B.* **2006**, *110*, 1727–1732.

[29] Zhuang, X.; Miranda, P. B.; Kim, D.; Shen, Y. R. *Phys. Rev. B* **1999**, *59*, 12632–12640.

[30] Bain, C. D.; Davies, P. B.; Ong, T. H.; Ward, R. N.; Brown, M. A. *Langmuir* **1991**, *7*, 1563–1566.

[31] Jayathilake, H. D.; Driscoll, J. A.; Bordenyuk, A. N.; Wu, L.; da Rocha, S. R. P.; Verani, C. N.; Benderskii, A. V. *Langmuir* **2009**, *25*, 6880–6886.

[32] Cecchet, F.; Lis, D.; amd Benoît Champagne, J. G.; Caudano, Y.; Silien, C.; Mani, A. A.; Thiry, P. A.; Peremans, A. *Chem. Phys. Chem.* **2010**, *11*, 607–615.

[33] Wang, H.-F. *Prog. Surf. Sci.* **2016**, *91*, 155–182.

[34] Wang, H.-F.; Velarde, L.; Gan, W.; Fu, L. *Annu. Rev. Phys. Chem.* **2015**, *66*, 189–216.

[35] Ohno, P. E.; Wang, H.-f.; Geiger, F. M. *Nature Comm.* **2017**, *8*, 1032.

[36] Ohno, P. E.; Wang, H.-F.; Paesani, F.; Skinner, J. L.; Geiger, F. M. *J. Phys. Chem. A* **2018**, *122*, 4457–4464.

[37] Hofmann, M. J.; Koelsch, P. *J. Chem. Phys.* **2015**, *143*, 134112.

[38] He, Y.; Wang, Y.; Wang, J.; Guo, W.; Wang, Z. *Opt. Lett.* **2016**, *41*, 874–877.

[39] Busson, B.; Tadjeddine, A. *J. Phys. Chem. C* **2009,** *113,* 21895–21902.

[40] Shen, Y. R. *J. Phys. Chem. C* **2012,** *116,* 15505–15509.

[41] Bell, G. R.; Bain, C. D.; Ward, R. N. *J. Chem. Soc. Faraday Trans.* **1996,** *92,* 515–523.

[42] Gan, W.; Wu, B.-H.; Zhang, Z.; Guo, Y.; Wang, H.-F. *J. Phys. Chem. C* **2007,** *111,* 8716–8725.

[43] Holman, J.; Davies, P. B.; Neivandt, D. J. *J. Phys. Chem. B* **2004,** *108,* 1396–1404.

[44] Ward, R. N.; Davies, P. B.; Bain, C. D. *J. Phys. Chem.* **1993,** *97,* 7141–7143.

[45] Holman, J.; Neivandt, D. J.; Davies, P. B. *Chem. Phys. Lett.* **2004,** *386,* 60–64.

[46] Covert, P. A.; FitzGerald, W. A.; Hore, D. K. *J. Chem. Phys.* **2012,** *137,* 014201.

[47] Hines, M. A.; Todd, J. A.; Guyot-Sionnest, P. *Langmuir* **1995,** *11,* 493–497.

[48] Dreesen, L.; Humbert, C.; Celebi, M.; Lemaire, J. J.; Mani, A. A.; Thiry, P. A.; Peremans, A. *Appl. Phys. B* **2002,** *74,* 621–625.

[49] Hu, P.; Li, X.; Li, B.; Han, X.; Zhang, F.; Chou, K.; Chen, Z.; Lu, X. *J. Phys. Chem. Lett.* **2018,** *9,* 5167–5172.

[50] Maia, F.; Miranda, P. *J. Phys. Chem. C* **2015,** *119,* 7386–7399.

[51] Jacob, J.; Lee, T.; Baldelli, S. *J. Phys. Chem. C* **2014,** *118,* 29126–29134.

[52] Nihonyanagi, S.; Sayama, A.; ; Ohshima, Y.; Tahara, T. *Chem. Lett.* **2019,** *48,* 1387–1390.

[53] Hirose, C.; Akamatsu, N.; Domen, K. *Appl. Spectrosc.* **1992,** *46,* 1051–1072.

[54] Wang, H.-F.; Gan, W.; Lu, R.; Rao, Y.; Wu, B. H. *Int. Rev. Phys. Chem.* **2005,** *24,* 191–256.

[55] Rao, Y.; Tao, Y.-S.; Wang, H. *J. Chem. Phys.* **2003**, *119*, 5226–5236.

[56] Gan, W.; Wu, D.; Zhang, Z.; Feng, R.-R.; Wang, H.-F. *J. Chem. Phys.* **2006**, *124*, 114705.

[57] Byrnesa, S. J.; Geisslerb, P. L.; Shen, Y. R. *Chem. Phys. Lett.* **2011**, *516*, 115–124.

[58] Hall, S. A.; Hickey, A. D.; Hore, D. K. *J. Phys. Chem. C* **2010**, *114*, 9748–9757.

[59] Jena, K. C.; Covert, P. A.; Hall, S. A.; Hore, D. K. *J. Phys. Chem. C* **2011**, *115*, 15570–15574.

[60] Hore, D. K.; Beaman, D. K.; Richmond, G. L. *J. Am. Chem. Soc.* **2005**, *127*, 9356–9357.

[61] Hore, D. K.; Beaman, D. K.; Parks, D. H.; Richmond, G. L. *J. Phys. Chem. B* **2005**, *109*, 16846–16851.

[62] Hall, S. A.; Jena, K. C.; Trudeau, T. G.; Hore, D. K. *J. Phys. Chem. C* **2011**, *115*, 11216–11225.

[63] Boyd, R. W. *Nonlinear Optics;* Academic Press: San Diego, 2nd ed.; 2003.

[64] Roke, S.; Kleyn, A. W.; Bonn, M. *Surf. Sci.* **2005**, *593*, 79–88.

[65] Roke, S.; Kleyn, A. W.; Bonn, M. *Chem. Phys. Lett.* **2003**, *370*, 227–232.

[66] Laaser, J. E.; Skoff, D. R.; Ho, J.-J.; Joo, Y.; Serrano, A. L.; Steinkruger, J. D.; Gopalan, P.; Gellman, S. H.; Zanni, M. T. *J. Am. Chem. Soc.* **2014**, *136*, 956–962.

[67] Zhang, B.; Tan, J.; Li, C.; Zhang, J.; Ye, S. *Langmuir* **2018**, *34*, 7554–7560.

[68] Wells, R. J. *J. Quant. Spectrosc. Rad. Transf.* **1999**, *62*, 29–48.

[69] Váczi, T. *Appl. Spectrosc.* **2014**, *68*, 1274–1278.

[70] Velarde, L.; Wang, H.-F. *Phys. Chem. Chem. Phys.* **2013,** *15,* 19970–19984.

[71] Velarde, L.; Zhang, X.; Lu, Z.; Joly, A. G.; Wang, Z.; Wang, H.-F. *J. Chem. Phys.* **2011,** *135,* 241102.

[72] Mifflin, A. L.; Velarde, L.; Ho, J.; Psciuk, B. T.; Negre, C. F. A.; Ebben, C. J.; Upshur, M. A.; Lu, Z.; Strick, B. L.; Thomson, R. J.; Batista, V. S.; Wang, H.-F.; Geiger, F. M. *J. Phys. Chem. A* **2015,** *119,* 1292–1302.

[73] Nihonyanagi, S.; Mondal, J. A.; Yamaguchi, S.; Tahara, T. *Annu. Rev. Phys. Chem.* **2013,** *64,* 579–603.

[74] Shen, Y. R. *Annu. Rev. Phys. Chem.* **2013,** *64,* 129–150.

[75] Jena, K.; Hung, K.-K.; Schwantje, T.; Hore, D. K. *J. Chem. Phys.* **2011,** *135,* 044704.

[76] Marquardt, D. *SIAM J. Appl. Math.* **1963,** *11,* 431–441.

[77] Fang, M.; Baldelli, S. *J. Phys. Chem. C* **2017,** *121,* 1591–1601.

[78] Dembro, R.; Steihaug, T. *Mathematical Programming* **1983,** *26,* 190–212.

[79] Zhang, Z.; Guo, Y.; Lu, Z.; Velarde, L.; Wang, H.-F. *J. Phys. Chem. C* **2012,** *116,* 2976–2987.

[80] Yang, W.-C.; Hore, D. K. *J. Chem. Phys.* **2018,** *149,* 174703.

[81] Stolle, R.; Marowsky, G.; Schwarzberg, E.; Berkovic, G. *Appl. Phys. B* **1996,** *63,* 491–498.

[82] Superfine, R.; Huang, J. Y.; Shen, Y. R. *Opt. Lett.* **1990,** *15,* 1276–1278.

[83] Shen, Y. R.; Ostroverkhov, V. *Chem. Rev.* **2006,** *106,* 1140–1154.

[84] Lu, R.; Rao, Y.; Zhang, W.-K.; Wang, H.-F. *Proc. SPIE Conf. Nonlinear. Spectrosc.* **2002,** *4812–4815,* 115–124.

[85] Jena, K. C.; Covert, P. A.; Hore, D. K. *J. Chem. Phys.* **2011,** *134,* 044712.

[86] Ji, N.; Ostroverkhov, V.; Chen, C.; Shen, Y. R. *J. Am. Chem. Soc.* **2007,** *129,* 10056–10057.

[87] Covert, P. A.; Hore, D. K. *J. Phys. Chem. C* **2015,** *119,* 271–276.

[88] Hutchings, D. C.; Sheik-Bahae, M.; Hagan, D. J.; van Stryland, E. W. *Opt. Quant. Electron.* **1992,** *24,* 1–30.

[89] Vartiainen, E. M.; Peiponen, K.-E.; Asakura, T. *Appl. Opt.* **1993,** *32,* 1126–1129.

[90] Bassani, F.; Scandolo, S. *Phys. Rev. B* **1991,** *44,* 8446–8453.

[91] Ogilvia, J. F.; Fee, G. J. *Commun. Math. Comput. Chem.* **2013,** *69,* 249–262.

[92] Brée, C.; Demircan, A.; Steinmeyer, G. *Phys. Rev. A* **2012,** *85,* 033806.

[93] Lucarini, V.; Peiponen, K.-E. *J. Chem. Phys.* **2003,** *119,* 620–627.

[94] Lucarini, V.; Saarinen, J. J.; Peiponen, K.-E. *J. Chem. Phys.* **2003,** *119,* 11095–11098.

[95] de Beer, A. G. F.; Samson, J.-S.; Hua, W.; Huang, Z.; Chen, X.; Allen, H. C.; Roke, S. *J. Chem. Phys.* **2011,** *135,* 224701.

[96] Johansson, P. K.; Koelsch, P. *J. Am. Chem. Soc.* **2014,** *136,* 13598–13601.

[97] Sovago, M.; Vartiainen, E.; Bonn, M. *J. Phys. Chem. C* **2009,** *113,* 6100–6106.

[98] Yang, P.-K.; Huang, J. Y. *J. Opt. Soc. Am. B* **1997,** *14,* 2443–2448.

[99] Yang, P.-K.; Huang, Y. Y. *J. Opt. Soc. Am. B* **2000,** *17,* 1216–1222.

[100] Tompkins, H. G.; Irene, E. A. *Handbook of Ellipsometry;* William Andrew, Inc.: Norwich, NY, 2005.

[101] Liu, W.; Tkatchenko, A.; Scheffler, M. *Acc. Chem. Res.* **2014,** *47,* 3369–3377.

[102] Casford, M. T.; Davies, P. B. *ACS Appl. Mater. Chem.* **2009,** *8,* 1672–1681.

[103] Duffy, D. C.; Friedmann, A.; Boggis, S. A.; Klenerman, D. *Langmuir* **1998,** *14,* 6518–6527.

[104] Chen, X.; Clarke, M. L.; Wang, J.; Chen, Z. *International Journal of Modern Physics B* **2005,** *19,* 691-713.

[105] Pogrzeba, T.; Schmidt, M.; Milojevic, N.; Urban, C.; Illner, M.; Repke, J.-U.; Schomäcker, R. *Ind. Eng. Chem. Res.* **2017,** *56,* 9934–9941.

[106] Han, Y.; Raghunathan, V.; ran Feng, R.; Maekawa, H.; Chung, C.-Y.; Feng, Y.; Potma, E. O.; Ge, N.-H. *J. Phys. Chem. B* **2013,** *117,* 6149–6156.

[107] Schwarzberg, E.; Berkovic, G.; Marowsky, G. *Appl. Phys. A* **1994,** *59,* 631–637.

[108] Chang, R. K.; Ducuing, J.; Bloembergen, N. *Phys. Rev. Lett.* **1965,** *15,* 6–8.

[109] Carriles, R.; An, Y. Q.; Downer, M. C. *Phys. Stat. Sol.* **2005,** *242,* 3001–3006.

[110] Chen, J.; Machida, S.; Yamanoto, Y. *Opt. Lett.* **1998,** *23,* 676–678.

[111] Kajikawa, K.; Wang, L.-M.; Isoshima, T.; Wada, T.; Knoll, W.; Sasabe, H.; Okada, S.; Nakanishi, H. *Thin Solid Films* **1996,** *284–285,* 612–614.

[112] Mifflin, A.; Musorrafiti, M.; Konek, C.; Geiger, F. *J. Phys. Chem. B* **2005,** *109,* 24386–24390.

[113] Wang, J.; Bisson, P. J.; Marmolejos, J. M.; Shultz, M. J. *J. Phys. Chem. Lett.* **2016,** *7,* 1945–1949.

[114] Yamaguchi, S. *J. Chem. Phys.* **2015**, *143*, 034202.

[115] Rich, C. C.; Lindberg, K. A.; Krummel, A. T. *J. Phys. Chem. Lett.* **2017**, *8*, 1331–1337.

[116] Weidner, T.; Apte, J. S.; Gamble, L. J.; Castner, D. G. *Langmuir* **2009**, *26*, 3433–3440.

[117] Cimatu, K.; Baldelli, S. *J. Am. Chem. Soc.* **2008**, *130*, 8030–8037.

[118] Zhang, H.; Romero, C.; Baldelli, S. *J. Phys. Chem. B* **2005**, *109*, 15520–15530.

[119] Miyamae, T. Probing Metal/Organic Interfaces Using Doubly-Resonant Sum Freuqency Generation Vibrational Spectroscoy. In *Vibrational Spectorscopy*, Vol. 5; de Caro, D., Ed.; InTech: Rijeka, 2012.

[120] Jang, J. H.; Lydiatt, F.; Lindsay, R.; Baldelli, S. *J. Phys. Chem. A* **2013**, *117*, 6288–6302.

[121] Liu, Y.; Wolf, L. K.; Messmer, M. C. *Langmuir* **2001**, *17*, 4329–4335.

[122] Bain, C. D.; Troughton, E. B.; Tao, Y.-T.; Evall, J.; Whitesides, G. M.; Nuzzo, R. G. . *J. Am. Chem. Soc.* **1989**, *111*, 321–335.

[123] Yang, W.-C.; Hore, D. K. *J. Phys. Chem. C* **2017**, *121*, 28043–28050.

[124] Breen, N. F.; Weidner, T.; Li, K.; Castner, D. G.; Drobny, G. P. *J. Am. Chem. Soc.* **2009**, *131*, 14148–14149.

[125] Weidner, T.; Breen, N. F.; Li, K.; Drobny, G. P.; Castner, D. G. *Proc. Natl. Acad. Sci. U.S.A.* **2010**, *107*, 13288–13293.

[126] Buck, M.; Eisert, F.; Grunze, M.; Trager, F. *Appl. Phys. A.* **1995**, *60*, 1–12.

[127] Shen, Y. R. Surface Spectroscopy by Nonlinear Optics. In *Proc. Int. School of Physics, Ernico Fermi*; Hansch, T. W.; Inguscio, M., Eds.; North Holland: Amsterdam, 1994.

[128] Morita, A. *Theory of Sum Frequency Generation Spectroscopy;* Springer: Singapore, 2018.

[129] Miranda, P. B.; Shen, Y. R. *J. Phys. Chem. B* **1999,** *103,* 3292–3307.

[130] Lu, R.; Gan, W.; Wang, H. *Chin. Sci. Bull.* **2003,** *48,* 2183–2187.

[131] Gan, W.; Wu, B.-h.; Chen, H.; Guo, Y.; Wang, H. *Chem. Phys. Letts.* **2005,** *406,* 467–473.

[132] Chen, H.; Gan, W.; Wu, B.-H.; Wu, D.; Zhang, Z.; Wang, H.-F. *Chem. Phys. Lett.* **2005,** *408,* 284–289.

[133] Hung, K.-K.; Stege, U.; Hore, D. K. *Appl. Spectrosc. Rev.* **2015,** *50,* 351–376.

[134] Wu, H.; Zhang, W.; Gan, W.; Cui, Z.; Wang, H. F. *J. Chem. Phys.* **2006,** *125,* 133203.

[135] Richter, L. J.; Yang, C. S. C.; Wilson, P. T.; Hacker, C. A.; van Zee, R. D.; Stapleton, J. J.; Allara, D. L.; Yao, Y.; Tour, J. M. *J. Phys. Chem. B* **2004,** *108,* 12547–12559.

[136] Pouthier, V.; Ramseyer, C.; Girardet, C. *J. Chem. Phys.* **1998,** *108,* 6502–6512.

[137] Siltanen, M.; Cattaneo, S.; Vuorimaa, E.; Lemmetyinen, H.; Katz, T. J.; Phillips, K. E. S.; Kauranen, M. *J. Chem. Phys.* **2004,** *121,* 1–4.

[138] Siltanen, M.; Kauranen, M. *Opt. Comm.* **2006,** *261,* 359–367.

[139] Busson, B.; Dalstein, L. *J. Chem. Phys.* **2018,** *149,* 034701.

[140] Busson, B.; Dalstein, L. *J. Chem. Phys.* **2018,** *149,* 154701.

[141] Bloembergen, N.; Pershan, P. S. *Phys. Rev.* **1962,** *128,* 606–622.

[142] Bloembergen, N.; Chang, R. K.; Lee, C. H. *Phys. Rev. Lett.* **1966,** *16,* 986–989.

[143] Brown, F.; Parks, R. E.; Sleeper, A. M. *Phys. Rev. Lett.* **1965,** *14,* 1029–1031.

[144] Baldelli, S. *Acc. Chem. Res.* **2008,** *41,* 421–431.

[145] Baldelli, S. *J. Phys. Chem. B* **2005,** *109,* 13049–13051.

[146] Baldelli, S.; Markovic, N.; Ross, P.; Shen, Y.-R.; Somorjai, G. *J. Phys. Chem. B* **1999,** *103,* 8920–8925.

[147] Tong, Y.; Lapointe, F.; Thämer, M.; Wolf, M.; Campen, R. K. *Angew. Chem. Int. Ed.* **2007,** *56,* 4211–4214.

[148] Guyot-Sionnest, P. *Chem. Phys. Lett.* **1990,** *172,* 341–345.

[149] Somorjai, G. A.; McCrea, K. R. *Adv. Catalysis* **2000,** *45,* 385–438.

[150] Dalstein, L.; Revel, A.; Humbert, C.; Busson, B. *J. Chem. Phys.* **2018,** *148,* 134701.

[151] van den Berg, S. A.; van der Ham, E. W. M.; Vrehen, Q. H. F.; Eliel, E. R. *Opt. Lett.* **1998,** *23,* 906–908.

[152] van der Ham, E. W. M.; Vrehen, Q. H. F.; Eliel, E. R.; Yakovlev, V. A.; Alieva, E. V.; Kuzik, L. A.; Petrov, J. E.; Sychugov, V. A.; van der Meer, A. F. G. *J. Opt. Soc. Am. B* **1999,** *16,* 1146–1152.

[153] Heinz, T. F. Second-order Nonlinear Optical Effects at Surfaces and Interfaces. In *Nonlinear Surface Electromagnetic Phenomena*; Ponath, H. E.; Stegeman, G. I., Eds.; North Holland: Amsterdam, 1991.

[154] Petukhov, A. V. *Phys. Rev. B* **1995,** *52,* 16901–16911.

[155] Maytorena, J. A.; Mendoza, B. S.; Mochán, W. L. *Phys. Rev. B* **1998,** *57,* 2569–2579.

[156] Mizrahi, V.; Sipe, J. E. *J. Opt. Soc. Am. B* **1988,** *5,* 660–667.

[157] Matranga, C.; Guyot-Sionnest, P. *J. Chem. Phys.* **2001,** *115,* 9503–9512.

[158] Yang, W.-C.; Busson, B.; Hore, D. K. *J. Chem. Phys.* **2019,** *152,* 084708.

[159] Belkin, M. A.; Kulakov, T. A.; Ernst, K.-H.; Yan, L.; Shen, Y. R. *Phys. Rev. Lett.* **2000,** *85,* 4474–4477.

[160] Yan, E. C. Y.; Fu, L.; Wang, Z.; Liu, W. *Chem. Rev.* **2014,** *114,* 8471–8498.

[161] Kauranen, M.; Cattaneo, S. *Prog. Opt.* **2008,** *51,* 69–120.

[162] Maytorena, J. A.; Mochán, W. L.; Mendoza, B. S. *Phys. Rev. B* **1995,** *51,* 2556–2562.

[163] Yamaguchi, S.; Shiratori, K.; Morita, A.; Tahara, T. *J. Chem. Phys.* **2011,** *134,* 184705.

[164] Sun, S.; Tian, C.; Shen, Y. R. *Proc. Nat. Acad. Sci. USA* **2015,** *112,* 5883–5887.

[165] Held, H.; Lvovsky, A. I.; Wei, X.; Shen, Y. R. *Phys. Rev. B* **2002,** *66,* 205110.

[166] Maikhuri, D.; Rurohit, S. P.; Mathur, K. C. *AIP Adv.* **2015,** *5,* 047115.

[167] Wei, X.; Hong, S.-C.; Lvovsky, A. I.; Held, H.; Shen, Y. R. *J. Phys. Chem. B* **2000,** *104,* 3349–3354.

[168] Zheng, R.; Weo, W. M.; Shi, Q. *Phys. Chem. Chem. Phys.* **2015,** *17,* 9068–9073.

[169] Guyot-Sionnest, P.; Chen, W.; Shen, Y. R. *Phys. Rev. B* **1986,** *33,* 8254–8263.

[170] Wang, F. X.; Rodríguez, F. J.; Albers, W. M.; Ahorinta, R.; Sipe, J. E.; Kauranen, M. *Phys. Rev. B* **2009,** *80,* 233402.

[171] Sipe, J. E.; Mizrahi, V.; Stegeman, G. I. *Phys. Rev. B* **1987,** *35*, 9091–9094.

[172] Guyot-Sionnest, P.; Shen, Y. R. *Phys. Rev. B* **1988,** *38*, 7985–7989.

[173] Mendoza, B. S.; Mochán, W. L.; Maytorena, J. A. *Phys. Rev. B* **1999,** *60*, 14334–14340.

[174] Jena, K. C.; Hore, D. K. *J. Phys. Chem. C* **2009,** *113*, 15364–15372.

[175] Sun, S.; Liang, R.; Xu, X.; Zhu, H.; Shen, Y. R.; Tian, C. *J. Chem. Phys.* **2016,** *144*, 244711.

[176] Fu, L.; Chen, S.-L.; Wang, H.-F. *J. Phys. Chem. B* **2016,** *120*, 1579–1589.

[177] Thämer, M.; Garling, T.; Campen, R. K.; Wolf, M. *J. Chem. Phys.* **2019,** *151*, 064707.

[178] Braun, R.; Casson, B. D.; Bain, C. D.; van der Ham, E. W. M.; Vrehen, Q. H. F.; Eliel, E. R.; Briggs, A. M.; Davies, P. B. *J. Chem. Phys.* **1999,** *110*, 4634–4640.

[179] Gao, X.; Liu, S.; Lu, H.; Gao, F.; Ma, H. *Ind. Eng. Chem. Res.* **2015,** *54*, 1941–1952.

[180] D.Karsa,, Ed.; *Proceedings of the Meeting on the Industrial Applications of Surfactants IV,* ; The Royal Society of Chemistry: Cambridge, UK, 1999.

[181] Tawfik, S.; Negm, N. *J. Mol. Liq.* **2016,** *215*, 185–196.

[182] Arora, P.; Singh, R.; Seshadri, G.; Tyagi, A. *tenside Surf. Det.* **2018,** *55*, 266–272.

[183] Liu, L.; Cao, T.; Zhang, Q.; Cui, C. *Adv. Mater. Sci. Eng.* **2018,** *None*, 9.

[184] Bentiss, F.; Traisnel, M.; Lagrenee, M. *J. Appl. Electrochem.* **2001,** *31*, 41–48.

[185] Moretti, G.; Quartarone, G.; Tassan, A.; Zingales, A. *Electrochim. Acta* **1996,** *41*, 1971–1980.

[186] Bentiss, F.; Lagrenee, M.; Traisnel, M.; Hornez, J. *Corrosion* **1999,** *55,* 968–976.

[187] Algaber, A.; El-Nemma, E.; M.Saleh, *Mater. Chem. Phys.* **2004,** *86,* 26–32.

[188] Migahed, M.; Azzam, E.; Al-Sabagh, A. *Mater. Chem. Phys.* **2004,** *85,* 273–279.

[189] Nahle, A.; Abu-Abdoun, I.; Abdel-Rahman, I. *Int. J. Corros* **2012,** *2012,* 380329–None.

[190] Labja, N.; Bentiss, F.; Lebrini, M.; Jama, C.; El-hajjaji, S. *Int. J. Corros* **2011,** *2011,* 548528–None.

[191] Liu, Z.; Jackson, T. Understanding Thermal Stability and Inhibition Effectiveness of Corrosion Inhibitors at High Temperatures. In ; CORROSION 2016 OnePetro: Vancouver, BC, 2016.

[192] Rossi, F.; Bevilacqua, M.; Busson, B.; Corva, M.; Tadjeddine, A.; Vizza, F.; Vesselli, E.; Bozzini, B. *J. Electroanal. Chem.* **2019,** *855,* 113641.

[193] Hosseinpour, S.; Hedberg, J.; Baldelli, S.; Leygraf, C.; Johnson, M. *J. Phys. Chem. C* **2011,** *115,* 23871–23879.

[194] Fockaert, L.; Ganzinga-Jurg, D.; Versluis, J.; Boelen, B.; Bakker, H.; Terryn, H.; Mol, J. *J. Phys. Chem. C* **2020,** *124,* 7127–7138.

[195] Holman, J.; Davies, P. B.; Nishida, T.; Ye, S.; Neivandt, D. J. *J. Phys. Chem. B* **2005,** *109,* 18723–18732.

[196] den Boer, J. H. W. G.; Kroesen, G. M. W.; de Hoog, F. J. *Meas. Sci. Technol.* **1997,** *8,* 484–492.

[197] Guo, W.; Chen, S.; Feng, Y.; Yang, C. *J. Phys. Chem. C* **2007,** *111,* 3109–3115.

[198] Lasia, A. *Electrochemical Impedance Spectroscopy and its Applications;* Springer Science Business Media: New York, USA, 2014.

[199] Mansfeld, F. *Corrosion* **1981,** *37,* 301–307.

[200] Mccafferty, E. *Corros. Sci.* **1997,** *39,* 243–254.

[201] Neves, R.; Robertis, E. D.; Motheo, A. *Appl. Surf. Sci* **2006,** *253,* 1379–1386.

[202] Cordoba-Torres, P. *Electrochim. Acta* **2017,** *225,* 592–604.

[203] Barsoukov, E.; Macdonald, J. *Impedance Spectroscopy: Theory, Experiment, and Applications;* John Wiley & Sons: Hoboken, NJ, 2nd ed.; 2005.

[204] Orazem, M.; Tribollet, B. *Electrochemical Impedance Spectroscopy;* John Wiley & Sons: Hoboken, NJ, 2008.

[205] Brug, G.; van den Eeden, A.; Sluyters-Rehbach, M.; Sluyters, J. *J. Electroanal. Chem.* **1984,** *176,* 275–295.

[206] Stern, M.; Geary, A. *J. Electrochem. Soc.* **1957,** *104,* 56.

[207] King, A.; Birbilis, N.; Scully, J. *Electrochim. Acta* **2014,** *121,* 394–406.

[208] Chen, Z.; Xiaohui, J.; Cheng, Z.; Liao, Y.; Pu, Q.; Duan, M. *Mater. Corros.* **2019,** *70,* 820–837.

[209] Bredar, A.; Chown, A.; Burton, A.; Farnum, B. *Appl. Energy Mater.* **2020,** *3,* 66–98.

[210] Wickman, B.; Fanta, A. B.; Burrows, A.; Hellman, A.; Wagner, J. B.; Iandolo, B. *Sci. Rep.* **2017,** *7,* 40500–None.

[211] Malitson, I. H. *J. Opt. Soc. Am.* **1965,** *55,* 1205–1209.

[212] Querry, M. R. "Optical Constants of Minerals and Other materials from the millimeter to the ultraviolet", Technical Report 88009, Chemical Research, Development, and Engineering Center, 1987.

[213] Johnson, P.; Christy, R. *Phys. Rev. B* **1974,** *9,* 5056.

[214] Werner, W.; Glantschnig, K.; Ambrosch-Draxl, C. *J. Phys. Chem. Ref. Data* **2009,** *38,* 1013.

[215] Ordal, M.; Bell, R.; Alexander, R.; Newquist, L.; Querry, M. *Appl. Opt.* **1988,** *27,* 1203–1209.

[216] Goldstein, H. *Classical Mechanics;* Addison-Wesley: Reeding, MA, 1980.

[217] Roy, S.; Hung, K.-K.; Stege, U.; Hore, D. K. *Appl. Spectrosc. Rev.* **2013,** *49,* 233–248.

[218] Wei, X.; Hong, S.; Zhuang, X.; Goto, T.; Shen, Y. *Phys. Rev. E* **2000,** *62,* 5160.

[219] Wei, X. *PhD book,* book, Department of Physics, University of California, Berkeley, 2000.

9 783384 271754